Social and economic effects of petroleum development in non-OPEC developing countries

Synthesis report

Social and economic effects of petroleum development in non-OPEC developing countries
Synthesis report

Jon McLin

International Labour Office Geneva

ISBN 92-2-105505-1

First published 1986

The designations employed in ILO publications, which are in conformity with United Nations practice, and the presentation of material therein do not imply the expression of any opinion whatsoever on the part of the International Labour Office concerning the legal status of any country, area or territory or of its authorities, or concerning the delimitation of its frontiers.
The responsibility for opinions expressed in signed articles, studies and other contributions rests solely with their authors, and publication does not constitute an endorsement by the International Labour Office of the opinions expressed in them.
Reference to names of firms and commercial products and processes does not imply their endorsement by the International Labour Office, and any failure to mention a particular firm, commercial product or process is not a sign of disapproval.

ILO publications can be obtained through major booksellers or ILO local offices in many countries, or direct from ILO Publications, International Labour Office, CH-1211 Geneva 22, Switzerland. A catalogue or list of new publications will be sent free of charge from the above address.

Printed in Switzerland POP

Preface

In 1979 the International Labour Office undertook a project of research and consultation aimed at helping oil and gas producers in the non-OPEC developing countries (NODCs) to learn from experience – their own and other producers' – how to derive greater social and economic benefit from their hydrocarbon resources. The project was initiated and carried out according to requests made by the ILO's 24-member Petroleum Committee. The main stages in the evolution of the project, and their relation to dramatic changes occurring during this period in the oil and gas markets, are summarised in Chapter 1.

This report, as its title denotes, represents an effort to synthesise in one document and in an organised fashion the factual and analytical material generated during the lifetime of the project. The readers at whom it is aimed comprise, among others, persons in policy-making or other responsible positions in government institutions, state or private petroleum enterprises, and organisations of petroleum workers in the NODCs. It has been kept as concise as possible with their time constraints in mind. Those seeking fuller information will find it in the relevant country case studies, most of which are being published separately.

There is another reason for this brevity: the Synthesis Report, being itself a summary report of the project, does not contain a final chapter summarising the argument and stating the general conclusions arrived at; general conclusions applicable to the wide variety of countries and situations dealt with would be, moreover, either impossible to formulate or too banal to be useful. Therefore if one wishes to get the benefit of this report, there is no alternative to going through it. The reader who does so should find that the report constitutes a kind of annotated checklist of the social and economic concomitants of petroleum development. Given the variety of circumstances that exist in petroleum-producing countries, the relative importance of various items on the list will vary from country to country. But everyone with responsibilities in this area should be expected to benefit not only from the treatment of issues of priority interest in their own situation but also from the sensitisation to the full range of issues – some of which may in that particular situation still lie in the future – that a reading of the full report will provide.

Two points should be made by way of clarification. The first concerns the somewhat unusual geographical and functional scope of this study. Why the *non-OPEC, oil and gas-producing* developing countries – rather than other, more familiar categories, such as developing countries in general, or oil-importing developing countries? "OPEC" was in fact used in symbolic terms. What was excluded in practice was a category of developing countries with very abundant petroleum resources relative to their size, i.e. the so-called "oil economies". Most of these are in OPEC, and most OPEC members can be so described. Most, but not all, as our coverage of Ecuador indicates. The reasons for the exclusion are that the "oil economy" phenomenon presents very special characteristics not typical of most developing countries and that it had been copiously written about. As for the decision to limit the study to petroleum producers, rather than to countries conforming to some indicator of net petroleum trade status (oil importer, self-sufficient, small exporter), the purpose was that the examination should not be limited to the financial and macro-economic implications of such a net petroleum trading position but should also include the developmental impacts associated with the upstream petroleum activities themselves.

The second point concerns the timing of the report's preparation in relation to developments in the international markets for oil and gas. The report was written in draft form in the summer of 1985, then revised in the period November 1985-January 1986. During the latter period, and more specifically in December 1985-January 1986, oil prices fell dramatically, by amounts of the order of 50 per cent. While pressures tending in that direction were evident during the time of writing and are noted at various places in the text – especially in relation to the impact OPEC price and production policies could be expected to have on the distribution of exploration activity and of production as between NODCs and OPEC members – the report is not based on any explicit scenario regarding future oil prices. It is not a guide to appropriate policies for a low-oil-price world; neither is it invalidated by low prices. Changing price expectations, like varying national situations, represent one factor among others to be borne in mind as readers go through the exercise of relating the general points made in this report to the concrete and specific situations with which they have to deal.

In the future the ILO, in its own work programme as well as through its contacts with other concerned organisations, will be alert to needs and opportunities to follow up the project's findings in practical ways.

Finally, I would like to thank the Government of Norway for the financial support which made possible the project; Mr. J. von Muralt, Chief, Basic Industries and Transport Branch, and Mr. L.R. Kohler, Industrial Specialist (Petroleum), for the contribution they have made to the development and execution of the project; and, of course, the author, Mr. Jon McLin, for his management of the project during its second and final phase.

Raymond S. Milne
Director,
Sectoral Activities Department

Contents

List of figures

List of tables

Summary

The boom experienced by many oil-producing developing countries after the shocks of 1973 and 1979 was not always effectively converted into lasting economic development. In the expectation that petroleum exploration in the non-OPEC developing countries (NODCs) could accelerate in the 1980s, this ILO project, funded by the Government of Norway, analysed the experience of several developing country producers in search of lessons that could be applied to future petroleum development. Emphasis was put on the petroleum sector's contribution to employment generation and manpower development, industrialisation and the impacts of petroleum activities at the local and regional level. Less attention was paid to more powerful, but more obvious, effects such as those on the balance of payments and on government revenues.

The NODCs' share of world oil production (one-seventh), although low, is higher than their share of proven reserves; and in the early 1980s it was rising. Exploration activity in these countries did not increase substantially after the two oil shocks, as might have been expected. On the other hand, it has not declined as early or sharply in the 1980s as in some other groups of countries. Pre-drilling activities – acreage holdings, geophysical surveys, and so on – which do not involve heavy expenditures, have held up rather well, suggesting sustained interest in these countries as possible future "plays". The extent to which this potential is realised will depend heavily on price trends and OPEC production policy: if lower cost production is available in adequate volumes, the incentive to explore in the NODCs will be limited.

Several of the NODCs have natural gas reserves as large as or larger than their oil deposits. Although few of these enjoy realistic export prospects, in several countries they are being economically exploited locally, for power generation, as industrial fuel and as feedstock for petrochemical projects. Of the countries considered here, Brazil, China, Egypt, India, Malaysia, Pakistan, Thailand and Trinidad and Tobago are all using their natural gas resources in promising ways.

Case studies were done for a sample of NODCs – Brazil, China, Côte d'Ivoire, Egypt, India, Malaysia, Pakistan, Peru and Thailand – and for Ecuador, an OPEC member. Summaries of their petroleum experience reveal considerable variety in national circumstances and approach, as compared with each other and to a selected group of producers (Mexico, Nigeria, Trinidad and Tobago) where petroleum plays a more dominant role in the economy. In the latter group, the relative importance of the petroleum sector is so great that the main policy issues concern macro-economic effects. Although in static terms these are strongly positive, over time dynamic influences operate which may have significantly negative impacts. These include the influence of an overvalued currency on tradeable goods and the tendency for the administrative sector to grow at the expense of the productive sector. For most NODCs, however, these problems stemming from relative petroleum abundance do not arise. Instead it is micro-economic effects which are our main concern.

Relatively few jobs are generated directly by petroleum activity, but studies of indirect effects show that there is a significant employment multiplier. The direct jobs have a high skill content, and to a large extent these skills are transferable to other activities. Provisions for indigenisation – preferential hiring of nationals, and training programmes in support of that objective – have become commonplace in contracts and in industry practice. International petroleum companies rely overwhelmingly on national staff in most cases, and the larger NODC producers – such as Brazil, China, Egypt, India and Mexico – employ almost entirely local personnel as regular staff in their petroleum industries. Foreign individuals and firms are involved mainly on a contract basis. Elsewhere, foreign operators tend to dominate exploration activity – although often making extensive use of local staff – and other high technology operations, while local companies work their way up the learning curve.

Most NODCs are able to supply some of the materials and services that the industry requires – construction, transportation, catering – but relatively few are or will be able to provide highly specialised equipment and services on competitive terms. Contracts typically require the use of local sources, provided that quality is acceptable. The larger NODCs – Brazil, China, India, Mexico – now provide from local sources a high proportion of their needs. For Peru a detailed assessment is presented of changes between 1968 and 1983 in local capabilities for different kinds of operations.

Forward-linked industrial projects may be viable in countries with cheap natural gas feedstock, assured access to markets and/or substantial industrial sectors. But these projects are capital-intensive and often risky. Among the smaller countries, Malaysia, Thailand and Trinidad and Tobago have undertaken ambitious, natural gas-based industrial schemes that will be worth watching.

Attention by the authorities is required to ensure that the impact of petroleum activities on the localities and regions where they are situated is positive. Proper planning and adequate financial means are needed to build and pay for the physical and social infrastructure involved, in such a way that local residents will be better off and feel better off than before such activities were initiated. This challenge is heightened by the tendency of these areas to be located in special environments: deserts, jungles and coastal areas supporting offshore operations. In many cases the designation of a portion of petroleum-related revenues for use in these local areas may be an appropriate instrument for meeting this challenge.

Petroleum development brings in its wake a number of social issues. These are generated both by production activities (influx of outsiders, "boom town" effects, environmental threats, the impact on tribal peoples) and by consumption effects (the social impact of new kinds of commercial energy, changes in the level and distribution of income, rising expectations). These issues are sometimes neglected – as priority is understandably given to increasing output – until problems become acute. If instead they are given timely attention, they can usually be managed in such a way that the positive effects will predominate. In this respect the long lead times characteristic of petroleum projects can, if utilised properly for planning purposes, be turned to good advantage.

Approximate conversion factors used

Oil	*1 tonne (metric) crude oil = 7.3 barrels*	
	1 barrel/day = 50 tonnes/year	
	1 barrel = 159 litres	
	= .159 cubic metres	
Natural gas		
	One billion (10^9) cubic metres*	*= 35 billion cubic ft*
		= .035 trillion (10^{12}) cubic ft*
	One billion cubic metres per year	*= 100 million cubic ft/day*
Oil/gas equivalence		
	One billion cubic metres of natural gas = 900,000 tonnes of oil	
	One billion cubic metres per year = 17,800 barrels of oil per day	
	One billion barrels of oil = 5 trillion (10^{12}) cubic ft	

*"Billion" is employed in this volume as it is commonly used in the oil industry to mean one thousand million, and "trillion" means one million million.

Abbreviations

b/d	= *barrels per day*
b/doe	= *barrels per day of oil equivalent*
boe	= *barrels of oil equivalent*
mtoe	= *million tonnes of oil equivalent*
mtpa	= *million tonnes per annum*
CPE	= *centrally planned economy*
LNG	= *liquefied natural gas*
LPG	= *liquefied petroleum gases*
NGL	= *natural gas liquids*
OIDC	= *oil importing developing countries*
OPEC	= *Organisation of Petroleum Exporting Countries*
R/P	= *reserves/production ratio*
km	= *kilometres*

Maps

Oil kindles extraordinary emotions and hopes, since oil is, above all, a great temptation. It is the temptation of ease, wealth, strength, fortune, power. It is a filthy, foul-smelling liquid that squirts obligingly up into the air and falls back to earth as a rustling shower of money. To discover and possess the source of oil is to feel as if, after wandering long underground, you had suddenly stumbled upon royal treasure. Not only do you become rich but you are also visited by the mystical conviction that some higher power has looked upon you with the eye of grace and magnanimously elevated you above others, electing you its favourite. Many photographs preserve the moment when the first oil spurts from the well: people jumping for joy, falling into each other's arms, weeping. Oil creates the illusion of a completely changed life – life without work, life that comes free. Oil is a resource that anesthetises thought, blurs vision, corrupts. People from poor countries go around thinking, God, if only we had oil! The concept of oil expresses perfectly the eternal human dream of wealth achieved through lucky accident, through a kiss of fortune, and not through sweat and hard work. In this sense, oil is a fairy tale and, like every fairy tale, a bit of a lie. Ryszard Kapuściński, "Revolution", in *The New Yorker,* 4 March 1985.

Chapter 1

Introduction

This is the report of the principal findings of a project initiated by the ILO in 1979. At that time, as the world was experiencing the second "oil shock", and as the World Bank was strengthening its programme of assistance for petroleum exploration and development in developing countries, it seemed probable that the following decade would witness a substantial increase in "upstream" petroleum activities in the non-OPEC developing countries (NODCs). The incentives were clear for the governments concerned to find and develop indigenous energy resources that could substitute for expensive oil imports, while for the international oil companies the higher prices and prospect of supply shortages expected for the mid- to late 1980s gave a new interest to the potential of these regions.

By that date it was also clear that the tremendous petroleum wealth generated in the past, and especially in the decade of the 1970s, had not always and everywhere been managed in ways that optimised the contribution to national development. The direct contributions of petroleum production to the balance of payments – and hence to easing the energy constraint on economic growth – and to government revenues were of course substantial and clearly visible. However, there were also significant indirect or secondary effects of the petroleum sector on the economy and society which were neither as well understood, as carefully planned nor as uniformly positive. The petroleum boom of the 1970s and early 1980s was accompanied by innumerable stories of petroleum-generated wealth misspent in corruption, grandiose prestige projects, bloated defence budgets, and other socially and economically ill-advised channels. The ills attributed to the oil boom included accelerated inflation, regressive effects on income distribution, weakening of the agricultural sector, growing international indebtedness (paradoxically despite the greater foreign exchange earnings), and exacerbated problems of rural-urban migration.

In these circumstances there was a case for taking stock of the relevant experience of existing petroleum producers in order to see what lessons it would yield that could help countries participating in the expected expansion of petroleum activity in NODCs to avoid the mistakes of the past and to optimise the contribution of their petroleum resources to their general economic and social development. That is the rationale for the project. The ILO, a specialised agency of the United Nations system with the mandate to promote the labour and social aspects of development, undertook the study in response to requests for action by its Petroleum Committee, a tripartite body comprising government, employers' and workers' representatives from 24 countries, which include significant petroleum producers and consumers from most regions of the world.

The project has been generously supported financially by the Norwegian Government, which as a recent petroleum producer itself has grappled thoughtfully with the issues with which this study is concerned.

The first phase of the study was conducted by a consulting firm, Norconsult, which began work in 1981-82. Norconsult teams made visits to eight petroleum-producing developing countries, including two members of OPEC. These were supplemented by desk studies on three additional developing countries and on five industrialised countries with experience of petroleum production in remote and difficult areas (Canada, Norway, United Kingdom, United States, USSR). The report of this phase was published in October 1984 under the title *Social and economic effects of petroleum development programmes* (Norconsult, 1984).

During the period that intervened between the beginning of the project and the publication of the Norconsult report, several of the background conditions changed. The first and most significant of these was the state of the hydrocarbons markets, which had altered almost beyond recognition. In 1981 it still seemed to many well-informed observers of the industry that the conditions of stringency experienced in 1979, and the tendency towards rising prices associated with it, would probably recur by the mid-1980s and would likely continue into the next decade. By late 1984 shortage had been replaced by surplus, prices had fallen by about a quarter in terms of constant United States dollars, and a return to stringency seemed to have been pushed out into the 1990s or beyond. (After some ups and downs in 1985, the market again weakened dramatically in early 1986, as prices fell sharply.)

The second noteworthy development, which was partly a result of the first, was that the expected sharp increase in exploration activity in the NODCs was not happening. Trends in exploration activity in the developing countries are reviewed in the next chapter. The data are not always satisfactory and on points of detail are sometimes ambiguous, but they are clear on the main point that there has been no significant increase in the level of exploratory activity in the NODCs, either in absolute terms or relative to the levels of activity elsewhere.

Third, the outlook for some particular countries has also changed significantly. Brazil, Egypt and India are examples of countries where production and estimated reserves have risen rapidly in recent years, albeit from a relatively modest base. More recently still, there have been encouraging discoveries in Colombia, the Yemen Arab Republic and the Syrian Arab Republic. On the other hand, early results of exploration offshore China have been modest, and the oil prospects of the Côte d'Ivoire appear more limited than was hoped in the early 1980s.

Fourth, while *internationally traded* natural gas faces prospects in the near term as reduced as those for oil, there has been in several countries an increased interest in developing indigenous natural gas deposits for certain *internal markets*, as a substitute for imported oil or for locally produced oil that could otherwise be exported. This trend has made it imperative to pay due attention in the study to the natural gas sector, which previously might have received scant notice in relation to the NODCs because the capital intensity associated with it was presumed to make it unsuitable for wide-scale use there.

A major feature of the category of countries under review here is the modesty of their resource base. If one excepts China and Mexico, then the NODCs collectively account for only around 5 per cent of the world's proved reserves of oil and gas combined. It is true that they have been less intensively explored than some other countries; however, that is in part a consequence of judgements made about their geological potential. It is also true that in petroleum exploration surprises are always possible; but with ever more sophisticated geological information and analysis available they get less likely.

The point is that, barring surprises, the issue for the overwhelming majority of NODCs is *not* what it was for a handful of countries, most of them in OPEC, in the 1970s and early 1980s – a problem of how to manage an abundance of petroleum revenues stemming from high levels of exports, relative to the size of the producing economy, and from rapidly rising petroleum prices. Most of the NODCs have little prospect of having a resource base on that scale, especially if judged in relation to their populations. The few exceptions, such as Brunei or Oman, are countries where the modesty of the resource base is more than offset by the smallness of the population. For most NODCs the first priority is to generate or attract – in an increasingly competitive business climate – the investment required to find and develop their hydrocarbon resources. Once that goal is achieved, the relevant issue is not how to eliminate the negative but how to accentuate the positive – in the sense of integrating the petroleum sector into their general economies and deriving as much developmental benefit from it as possible. Given that the petroleum sector makes intensive use of those resources in which NODCs generally tend to be most deficient – capital and advanced technology – it is all the more important that resources devoted to petroleum development be used as effectively as possible.

This revised appreciation of the nature of the issue has influenced the research design of the project. In particular, it was decided to supplement the Norconsult report with a number of case studies of countries selected with the above assessment in mind. The group comprises Brazil, China, Côte d'Ivoire, Ecuador (an OPEC member but one with more in common with the NODCs than with most of the rest of OPEC), Egypt, India, Malaysia, Pakistan, Peru and Thailand. Most of the studies, which were carried out in late 1984 and 1985, were done by nationals or other long-term residents of the countries in question; most are being published by the ILO. While the countries as a group encompass considerable variety in other respects, they are not major oil exporters either in relation to the size of their economies or as factors in the world petroleum markets – although by one or both criteria China, Ecuador and Malaysia approach being so. They are an intermediate category, with enough of a resource base and a history of exploiting it to make a review of their experience worth while, but without the overwhelming dependence on petroleum that would make them atypical of the NODC group that we are concerned with here.

The following are the external collaborators who prepared reports on the countries indicated. Where not otherwise indicated in this study, information pertaining to these countries is taken from their papers, which are cited in full in the Bibliography.

Brazil	• Norman Gall
China	• Peter Hills
Ecuador	• Cornelio Marchán Carrasco
Egypt	• Yasser Sabra
India	• Tata Energy Research Institute, with C. R. Jagannathan
	• A. D. Saramma
Malaysia	• Paddy Bowie
Mexico	• Gustavo Rodríguez Elizarrarás
Peru	• Eleodoro Mayorga Alba
Thailand	• Charit Tingsabadh
	• Faculty of Economics, Chulalongkorn University
USSR	• A. D. Khitoun

The intended beneficiaries of the study are principally the NODCs whose petroleum experience lies largely in the future, either because they become significant producers for the first time or because they expand their production well beyond historical levels. National circumstances tend to be quite specific, but this study was conducted in the hope that there is none the less value in exchanging experience. In that spirit, the findings have been presented in consultations with government, industry and labour representatives from the petroleum sector in a number of NODCs with hopes of expanded petroleum output. The first of these were held in Bangladesh and Guatemala in October-November 1985.

Chapter 2

The petroleum upstream in the NODCs: Trends and characteristics

The scope for using petroleum development to pursue other economic and social goals depends in the first instance of course on the size of the resource base and on the scale of exploration, development and production activity. This chapter is therefore devoted to an examination of current estimates of the NODC resource base and a brief review of recent upstream activities in these countries.

The NODCs which have exploration and/or production activities are a very diverse group. At one extreme, they include Mexico, which alone accounts for half of the proved oil reserves of the entire group. At the other, there are approximately 40 developing countries which produce no petroleum but where some kind of exploration is taking place. The classification in figure 1 is useful for our purposes.

This classification usefully distinguishes between those countries which are major exporters – whether high or middle income, whether OPEC members or not – and those which are not. The ILO project is concerned mainly with the minor exporters and the producers who are net importers, and less so, for reasons specified in Chapter 1, with the major exporters. This characteristic, rather than the formal one of OPEC membership or non-membership, will define our field of inquiry. Accordingly, we will on occasion extract Mexico from the NODC figures, as its resource base puts it in the category of the major exporters and tends to have a distorting effect on collective figures for the NODC group.

Oil reserves and production

The resource base of the NODCs is modest, as has already been noted. In respect of oil they account for about one-seventh of the world's proved reserves and one-sixth of current production. The composition of that total is lumpy: Mexico accounts for half the group's reserves and nearly a third of its production. China contributes a further fifth of reserves and production. Over 30 NODC producers share the remaining 30 per cent of the group's modest collective reserves and the remaining 45 per cent of its production. Table 1 gives further details.

It is noteworthy that the NODCs' share of production is somewhat higher than their share of reserves. The reverse was true as of 1980 (Favre and Leleuch, 1982). This change reflects the recent tendency of the low cost Middle East producers, which account for a huge share of world reserves, to curtail their production as they functioned as residual suppliers to oversupplied international markets. Another manifestation of the tendency of the Gulf countries and other

Figure 1. Third World country classification

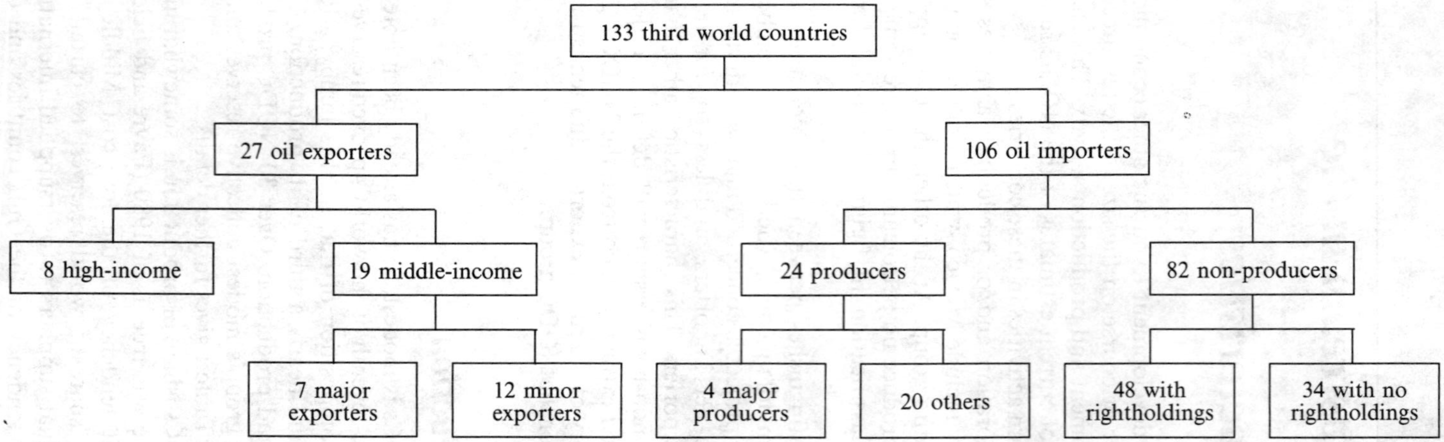

8 high-income	7 major exporters	12 minor exporters	4 major producers
Bahrain	Algeria*	Angola	Argentina
Brunei	Indonesia*	Cameroon	Brazil
Kuwait*	Islamic Republic of Iran*	China	Colombia
Libyan Arab Jamahiriya*	Iraq*	Congo	India
Oman	Mexico	Ecuador*	
Qatar*	Nigeria*	Egypt	
Saudi Arabia*	Venezuela*	Gabon*	
United Arab Emirates*		Malaysia	
		Peru	
		Syrian Arab Republic	
		Trinidad and Tobago	
		Tunisia	

* Member of OPEC

Source: John Foster, EDPRA Consulting Inc.: "Petroleum exploration and development trends in the developing countries", paper presented at the United Nations Symposium on Financing of Petroleum Exploration and Development in Developing Countries, Athens, April 1985.

Table 1. Oil reserves (published/proved) and production, 1984

	Reserves			Production		
	10^9 (tonnes)	10^9 (barrels)	Share of world total (%)	10^6 (tonnes /year)	10^3 (barrels /day)	Share of world total (%)
Latin America	11.7	83.3	11.8	338.0	6 705	11.9
Less OPEC	7.8	56.7	8.1	227.9	4 575	8.1
Less OPEC and Mexico	1.0	8.1	1.0	77.9	1 565	2.7
of which						
Argentina	0.3	2.3	0.3	24.3	465	0.9
Brazil	0.3	2.0	0.3	23.8	495	0.8
Middle East	54.2	398.4	56.4	577.7	11 735	20.3
Less OPEC	0.8	5.7	0.8	32.8	645	1.1
Africa	7.5	55.6	8.0	248.8	5 185	8.9
Less OPEC	1.2	8.8	1.2	74.9	1 520	2.6
of which						
Angola	0.3	1.8	0.3	10.3	207	0.3
Egypt	0.4	3.2	0.4	45.2	915	1.6
Tunisia	0.2	1.5	0.2	5.7	114	0.2
Asia/Australasia (excluding CPEs)	2.5	18.6	2.5	160.5	3 320	5.8
Less OPEC, Brunei, Japan, Australia, New Zealand	0.9	6.8	0.9	56.6	1 165	2.0
of which						
India	0.5	3.5	0.5	27.1	543	0.9
Malaysia	0.4	3.0	0.4	24.8	525	0.9
China	2.6	19.1	2.7	114.5	2 300	4.1
Total world	96.1	707.2	100.0	2 826.1	57 800	100.0
of which						
OPEC	64.8	476.3	67.2	899.8	18 345	31.8
NODCs	13.3	97.1	13.8	506.7	10 205	17.9
NODCs less Mexico	6.5	48.5	6.7	356.7	7 195	12.6

Source: Adapted from *BP Statistical Review of World Energy, 1985*, and other sources.

OPEC members to "underproduce" is the declining share of internationally traded oil in total world consumption, from 56 per cent in 1979 to 43 per cent in 1983 (ILO, 1986, I). The implication is that so long as OPEC continues to be subject to production constraints, additional reserves discovered or developed in the NODCs are more likely to pass into production than those in OPEC, a factor which will have its effect on the distribution of world-wide exploration activity. Furthermore, the combination of declining consumption of oil and declining share of consumption attributable to international trade has magnified the impact that modest NODC production levels have had on their self-sufficiency ratios.

The picture just described is not static. The share of world oil output from NODCs is rising. Of the approximately 30 NODCs producing oil in 1985, ten produced little or nothing in 1970 (Benin, Cameroon, Congo, Côte d'Ivoire, Ghana, Guatemala, Malaysia, Philippines, Thailand, Zaire). Their aggregate 1985 production was about 800,000 b/d, nearly 60 per cent of it accounted for by Malaysia. Another six NODCs (Brazil, Egypt, India, Mexico, Peru, Syrian Arab Republic) which were already producers in 1970 have expanded their production several-fold; for the first four of these, volumes have reached substantial levels. Although the bulk of the volume increase achieved by this group is accounted for by Mexico's spectacular rise (62 per cent of the increase), the other five all tripled or nearly tripled their output. The remaining 15 or so NODC producers have not changed their positions much since 1970, although some of them have undergone slight increases (Angola, Argentina, Oman, Syrian Arab Republic) or decreases (Turkey) in output (F. Parra, 1985).

Exploration

There is widespread agreement that the NODCs' share of the world-wide exploration effort in the past has been less than would be expected on the sole basis of their geological prospectivity. A number of recent studies have documented and analysed this situation (Blitzer et al., 1983; Broadman, 1985; Grossling and Neilsen, 1985). According to one count, in the year 1980 17,405 exploratory wells were drilled in the world outside the centrally planned economies; of those 16,278 were in developed countries (of which 15,773 in North America), 636 in oil-exporting developing countries, mainly OPEC, and 491 in oil-importing developing countries (OIDCs) (Blitzer et al., 1983). By another estimate, the United States has only 11 per cent of the world's prospective area but over the years has accounted for 71 per cent of the wells and 42 per cent of the seismic lines (Grossling and Neilsen, 1985).

There is disagreement about the extent to which this distribution of effort is disproportionate to the distribution of reserves; to some extent the "creaming" effect, whereby the best prospects are explored first, is certainly at work. But this debate is about matters of degree. There is also disagreement about the factors that are responsible for this pattern. "Institutional" explanations are certainly important. Many of the operators and rigs doing exploration in North America are not internationally mobile, and the fiscal and infrastructural context in which they operate is not available elsewhere. Exploration in many OPEC countries is economically unattractive given present reserves/production ratios, which are influenced by current pricing and production policies. Cost is hardly the explanation. The MIT study (Blitzer et al., 1983) concluded that in the ten-year period, 1972-81, 52.6 million barrels of oil and gas equivalent were added to reserves for each wildcat drilled in the oil-exporting developing countries, at an average cost per added barrel of US$0.24; in the oil-importing developing countries the figures were 5.2 million barrels and US$1.15; in the industrialised countries 0.7 million and US$4.48 (see *Petroleum Economist*, May 1985). Broadman's (1985) analysis leads to

***Figure 2.* Exploratory drilling (except United States, Canada and centrally planned economies)**

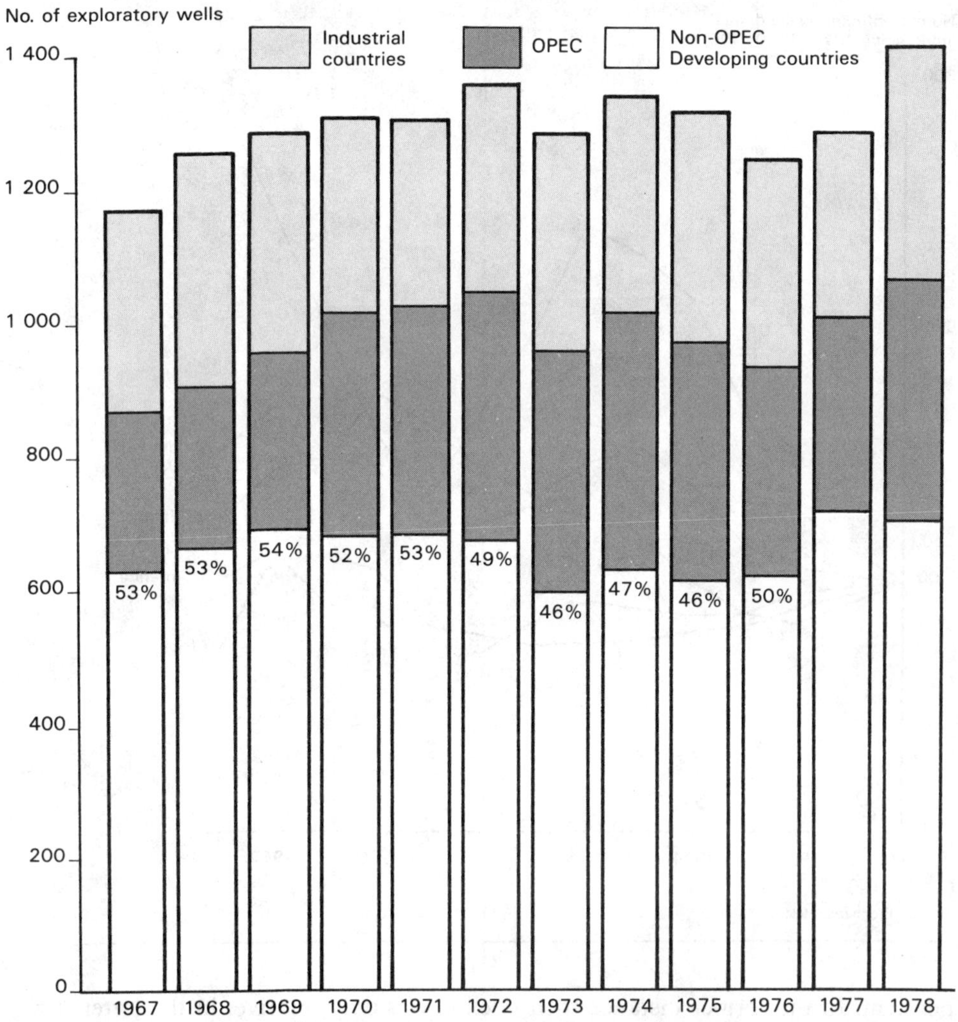

Source: Exxon, 1985.

the conclusion that contractual arrangements and taxes have a major influence on the distribution of drilling as among oil-producing areas, while geological potential is the dominant determinant of drilling patterns in areas which are not yet oil-producing.

To the surprise of many, exploration in the NODCs (or the OIDCs, to take a largely overlapping category) did not increase significantly in the years following the 1973 and 1979 price increases. The MIT data show that the OIDC share of exploratory wells drilled world-wide dropped from 3.37 per cent in 1972 to 2.82 per cent in 1980, and their share of seismic activity from 15.75 per cent to 8.10

Figure 3. Exploration drilling history

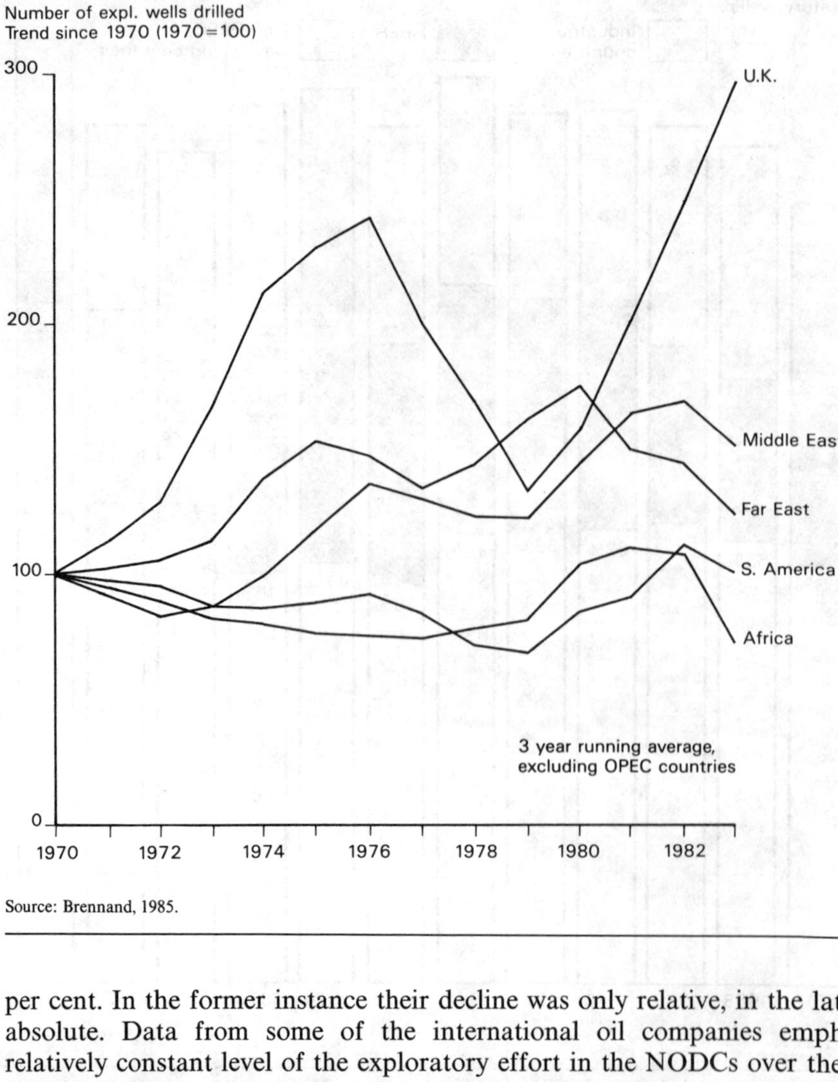

Source: Brennand, 1985.

per cent. In the former instance their decline was only relative, in the latter it was absolute. Data from some of the international oil companies emphasise the relatively constant level of the exploratory effort in the NODCs over the years, as measured in exploratory wells drilled (see figures 2 and 3).

Since the oil market peaked in 1980-81, it appears that exploratory drilling has declined in all NODC regions but that the decline began later and was less sharp than that experienced in the developed market economies (see table 2; these figures do not exclude OPEC countries, but that fact does not affect the conclusion, since the decline affected that group at least to the same extent). On the other hand, figure 4 shows that, as measured by licensed acreage, exploration activity has not diminished much in the NODCs since 1980 except in the Middle East. Some of the international oil companies, such as Amoco and Elf-Aquitaine, have directed a relatively large share of their recent exploration effort at countries in the NODC

Table 2. Drilling rigs active in the market economies, 1980-84, by quarter [1] (number of units)

	1980		1981				1982				1983				1984		
	3	4	1	2	3	4	1	2	3	4	1	2	3	4	1	2	3
Developed market economies	3 776	3 875	4 180	4 502	4 768	5 112	4 215	3 378	2 894	3 422	2 398	2 409	2 768	3 175	2 770	2 932	2 931
Canada	422	276	251	219	223	267	247	130	146	352	169	170	208	223	297	258	184
Europe	216	227	236	235	253	266	270	279	263	253	227	208	197	185	214	220	234
United States	3 099	3 337	3 656	4 007	4 242	4 530	3 639	2 908	2 429	2 768	1 956	1 986	2 312	2 710	2 206	2 401	2 452
Other	39	35	37	41	50	49	59	61	56	49	46	45	51	57	53	54	61
Developing countries	1 058	1 089	1 127	1 086	1 118	1 121	1 140	1 135	1 136	1 144	1 048	982	991	952	938	935	958
Africa	241	239	239	239	239	238	231	249	247	245	217	180	158	159	147	136	135
Asia	212	223	231	231	236	233	249	251	252	246	239	210	247	233	230	226	233
Middle East	128	128	139	107	113	114	119	124	126	154	132	141	143	135	129	141	136
Latin America	477	499	518	509	530	536	541	511	511	499	460	451	443	425	432	432	454

[1] As at the last day of a given quarter.

Source: Energy Economic Research Ltd.: *Oil and Energy Trends*, various issues, as reproduced in United Nations Committee on Natural Resources, 1985.

Figure 4. Concession history

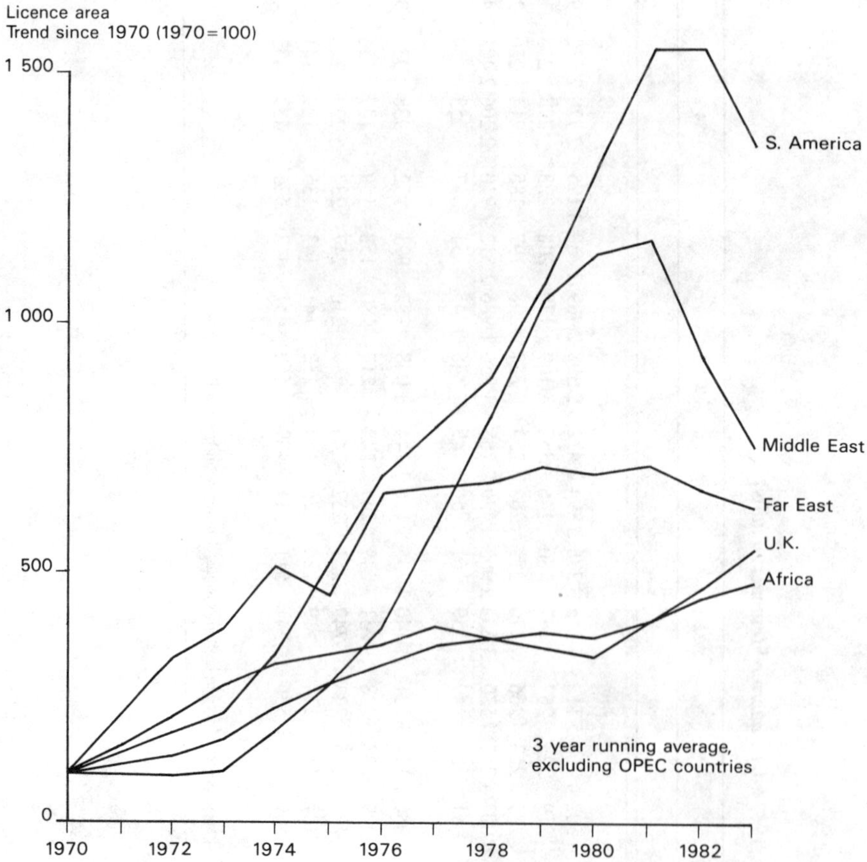

Source: Brennand, 1985.

group. A spokesman for the Royal Dutch/Shell Group pointed out that the proportion of its world-wide exploration expenditure going to developing countries where there were no existing large, established production operations doubled, from 20 per cent to 40 per cent, between 1979 and 1985 (van Engelshoven, 1985).

It may be surprising that this degree of interest, albeit limited, is shown for NODC areas under prevailing weak market conditions. Part of the explanation must be the continued effective exclusion of the commercial operators from many of the lower-cost OPEC areas, combined with the continuing depletion, rising costs and smaller finds in the industrialised countries. The bargaining power of the NODCs is not as great as it was a decade ago, when concern was running high about access to resources in a supposedly energy-short world. Many of them have revised their petroleum policies and legislation in order to make their terms more attractive to investors in this more competitive situation. But given companies' needs to think about their reserves position in the 1990s, the NODCs' position has not weakened as much as might have been expected.

It may be noteworthy that, according to some indications, a heavy share of recent exploration expenditure in the NODCs went to Africa (ILO, 1986, I). For example, one study which focuses specifically on exploratory activity in those NODCs which are not now producers, finds that a high proportion of acreage holdings, seismic activity and exploratory drilling in that group of 48 countries is concentrated in sub-Saharan Africa. The level of drilling activity, however, is not high, and is concentrated in the small number of countries that have had discoveries: "over the ten years 1972-81, 124 exploratory wells had been completed in 19 non-producing countries ... more than half these wells were in four countries with discoveries: Chad, Somalia, Sudan and Tanzania" (Foster, 1985).

This uneven distribution of exploratory activity is not limited to Africa but is a general characteristic of the NODCs. Through most of the 1970s around half of the exploratory wells drilled in all OIDCs were drilled in Argentina, Brazil and India by the national oil companies of those countries (World Bank, 1983a). That tendency of the large countries to dominate the totals continues. There results a somewhat misleading picture of the share of total activity accounted for by national oil companies. Whereas 66 per cent of the exploratory wells drilled in energy deficient developing countries in 1981 were drilled by national oil companies, that figure drops to 41 per cent if Argentina (YPF) and Brazil (Petrobrás) are excluded (United Nations, *Development and International Economic Co-operation*, 24 August 1984).

There is also a marked tendency for exploration to be concentrated in countries where hydrocarbons have already been found in commercial quantities.

In forming an overall judgement on these questions, one must recognise the tremendous data problems involved. Despite the numerous studies, it is difficult to ensure comparability of data, as they are compiled for different groups of countries, over different periods, using different measures of effort. Expenditure figures, which in some ways would be the most meaningful, are especially inadequate. With all the qualifications this situation imposes, what emerges from all of the above attempts at measuring trends is a judgement that the exploration effort in the NODCs has been remarkably level in *absolute* terms over the past 15 years or so, especially if measured by numbers of exploratory wells. In terms of the less expensive pre-drilling activities, there may have been an increase; this suggests that many operators are seeking to keep open the option of expanding their activities there, should market conditions in future warrant it, while minimising their current financial outlays. Assessing this effort in *relative* terms, compared with that in OPEC or in the industrial countries, is more difficult because activity in these other areas has varied more sharply as a function of various special factors: the particular features of the petroleum industry in North America, the timing of the development phase of the North Sea, and the "swing" role of OPEC in absorbing adjustments required by the overall state of the world oil market.

It is hazardous to speculate about the future prospects for upstream activity in the NODCs, as is shown by the dismal record of earlier predictions related to the petroleum industry. What may be appropriately said is that these prospects will depend critically on the related factors of price and OPEC production policy. To the extent that markets are influenced by an expectation of a firming of price trends in the 1990s – this implies continuing restraint by OPEC on its production, to the benefit of higher-cost suppliers – the NODC prospects may be expected to improve. If, on the other hand, there is a continuing erosion of the real price of petroleum, then upstream activity in the NODCs, which tend to be high-cost producers relative to OPEC, will be more constrained, although it will be protected against collapse by a continuation of government policies favoring indigenous production over imports.

Governments of NODCs face a daunting challenge in deciding in the face of these uncertainties on the nature and scope of the national petroleum industry that they wish to have. The uncertainty is magnified by the time scale applicable to the industry; while it is difficult to predict conditions in one year, investment decisions must typically be made a decade ahead. The scale of resources involved means that the penalty for getting it wrong is very high. But the choice cannot be escaped.

Natural gas

World reserves of natural gas are almost equal, in terms of thermal equivalence, to those of oil. The proportion is in the order of 90 per cent, although the figure for gas reserves is more uncertain, since relatively little exploration has taken place that is aimed specifically at finding gas. However, in production terms gas output amounts to only about half that of oil. The reason for the discrepancy, of course, lies mainly with the relatively greater expense and inconvenience involved in transporting gas, compared with oil. That factor is especially important in respect of international movements of fuel. Table 3 shows how the NODCs fit into the world gas scene.

The natural gas issues that are of most interest in relation to the NODCs concern gas that is utilised internally rather than exported. A much smaller fraction of gas than of oil is traded internationally in any case (respectively 5 per cent vs. 50 per cent moves inter-regionally; Hartshorn, 1985). Added to that, several factors have contributed recently to diminish the prospects for new natural gas export schemes for some years to come. Those with the greatest likelihood of being realised are based on reserves that are relatively large, accessible and cheap. Although some of the NODCs – such as Malaysia, Mexico, Trinidad and Tobago – may fit that category, most do not. However, numerous observers have recently pointed out the economic advantages open to such modestly endowed countries if they develop their natural gas resources for internal use (Gault, 1983, 1985; Bourcier, 1984; ILO, 1986, I; World Bank, 1983b). The arguments are based on points such as: lack of alternative use for associated gas that would otherwise be flared; the more modest investment outlay required, compared to export projects; and the closer integration with the local economy that characterises domestic projects. Some developing countries were among the more rapidly growing markets for natural gas in the first half of the 1980s, and this rapid growth was expected to

Table 3. Natural gas reserves (published/proved) and production,[1] 1984

	Reserves			Production		
	10⁶ tonnes oil equivalent	10^9 m³	Share of world total (%)	10⁶ tonnes oil equivalent /year	10^9 m³ /year	Share of world total (%)
Latin America	4 770	5 300	5.4	71.1	79.0	4.9
Less OPEC	3 240	3 600	3.7	56.2	62.4	3.8
Less OPEC and Mexico	1 260	1 400	1.4	24.1	26.7	1.6
of which						
Argentina	638	700	0.7	9.1	10.1	0.6
Trinidad and Tobago	270	300	0.3	2.8	3.1	0.2
Middle East	22 140	24 600	25.6	38.4	42.6	2.6
Less OPEC	810	900	0.9	9.6	10.6	0.6
Africa	4 770	5 300	5.5	26.6	29.5	1.9
Less OPEC	540	600	0.6	3.6	4.0	0.2
of which Egypt	180	200	0.2			
Asia/Australasia (excluding CPEs)	4 140	4 600	4.9	68.0	75.5	4.7
Less OPEC, Brunei, Japan, Australia, New Zealand	2 340	2 600	2.7	24.8	27.5	1.7
of which						
India	360	400	0.4			
Malaysia	1 260	1 400	1.5	3.7	4.1	0.2
Pakistan	360	400	1.5	12.1	13.4	0.8
China	810	900	0.9	11.5	12.7	0.8
USSR	36 990	41 100	42.7	526.5	579.2	36.5
Total world	86 580	96 200	100.0	1 444.4	1 604.8	100.0
of which						
OPEC	25 920	28 800	29.9	83.1	92.3	5.7
NODCs	7 740	8 600	8.9	105.7	117.4	7.3
NODCs less Mexico	5 760	6 400	6.6	73.1	81.2	5.0

[1] Excluding gas flared or recycled.
Source: Adapted from *BP Statistical Review of World Energy, 1985*, and other sources.

continue to the end of the century (*Petroleum Intelligence Weekly*, 4 November 1985).

Several of the NODCs have already moved to develop their natural gas resources for indigenous use. Some of these are noted in Chapter 3, notably Brazil, Côte d'Ivoire, Egypt, India, Pakistan, Thailand and Trinidad and Tobago. Elsewhere, Bangladesh has reduced its oil import bill to 45 per cent of its export earnings (from a high of 85 per cent), largely through a vigorous programme of substituting indigenous gas for imported oil. Output for 1985 was about 1.7 mtoe. Although most of it was used for power generation (37 per cent in 1983/84) and fertiliser manufacture (40 per cent), the remainder was used to fuel a variety of industrial plants and a small amount of residential and commercial consumption. Some of the residential/commercial distribution network is innovatively reaching the lower social strata, and the social effects are significant. Some of the industrial users, such as glass factories and brickworks, generated substantial employment; an estimated 150 gas-fuelled brickworks employed 300 or more people each. A separation plant was being introduced which would make possible the recovery of significant quantities of LPG, the availability of which will improve the standard of living and ease pressures on fuelwood in many parts of the country. The United Republic of Tanzania, which has limited immediate scope for substituting its gas resources for petroleum imports – since most of its petroleum products are used in the transportation sector – is seeking to use the gas resource to improve its balance of payments by another means. It is developing its offshore Songo Songo gas deposit with a view to building an ammonia and urea plant that would export 95 per cent of its output. In South America, Argentina, Bolivia, Chile and Colombia all have potential for greater utilisation of their domestic gas resources (*Petroleum Intelligence Weekly*, 4 November 1985).

Chapter 3

Main features of petroleum development: Some national cases

A great deal of empirical material has been generated by this project. It takes the form of commissioned papers, notes and reports of country visits, desk studies, and published books and articles. In order to utilise this data base as fully as possible within the confines of this brief report, the material will be synthesised in two ways. In this chapter national cases will be reviewed, highlighting the special characteristics, priorities and policy choices that have been prominent in our sample of countries. The purpose will be to identify and summarise those features of petroleum development which are most significant for the broader patterns of economic and social development in each country; as will be seen, these vary a great deal from one country to another. The remaining chapters approach the same set of data in an analytical way, looking in a cross-national manner at issues common to several countries.

The first group of countries presented here are those on which relatively complete and/or first-hand information was gathered. This set is reasonably representative of the NODCs as a whole. The second group of countries described are major exporters. These accounts are based on reviews of the literature rather than on commissioned studies, and they deal with cases that are not typical of most NODCs. However, their problems are those to which NODCs in a sense aspire, in that they are problems stemming from resource abundance. That is why they are included here.

Modest producers

Brazil

The distinguishing characteristic of Brazil's petroleum history has been the sense of high policy – verging on national mission – with which it has been invested from a very early date. A state oil monopoly was created in 1938, in a move inspired by the nationalisation of the petroleum industry in Mexico a month earlier; this was a year before any oil had been found in Brazil. Petrobrás, established in 1954 as the successor state oil company, has come to command vast resources, roughly equal in financial terms to those of all 540 other state companies in the country. In 1984 its gross revenues amounted to 9 per cent of GDP and exceeded those of the federal treasury. As it grew to this size it has been, simultaneously or consecutively, a or the main instrument for pursuing a number of major state policies not limited to the petroleum business.

These policies were at first based on strategic and nationalist considerations. They aimed at establishing the fuel supply sources and distribution networks required to create a national road and power infrastructure. There followed a period marked by heavy investment in refining, aimed at capturing value added, generating employment, favouring the development of certain regions, and stimulating the growth of a Brazilian capital goods industry. Conserving scarce foreign exchange by finding and developing indigenous hydrocarbon resources has consistently been a goal, but it took on a new priority and urgency following the oil price increases of the 1970s. Since that time exploration and development efforts have been intensified, especially offshore; this followed an earlier emphasis on onshore exploration in the Amazon basin, with generally disappointing results. The pace has been forced by a Petrobrás and government leadership, under the military regime, that was highly conscious of the economic stakes involved for the nation. In the 1980s the payoff has come, in the form of rapidly rising reserves and production figures. Oil production rose from 187,000 b/d in 1980 to 422,000 b/d in 1984; by mid-1985 production had risen to about 550,000 b/d and was accounting for some 60 per cent of national consumption. In early 1985 Petrobrás announced that an offshore well in the Campos basin, drilled in a water depth of 853 meters, had encountered what appeared to be a giant field with recoverable reserves on the order of 500 million barrels. This, together with two other Campos fields discovered in 1984, were thought to add over 1 billion barrels to the country's reserves.

Brazil may also be on the brink of significant gas exploitation. Substantial deposits have been found in the Amazon region, offshore São Paulo state (where the largest potential markets are located) and elsewhere. Development has so far been hesitant for several reasons: the multibillion dollar cost of the pipeline required to bring the Jurua gas (the Amazonian find) to market; competition from other fuels in certain markets (ample hydropower, surplus fuel oil because of refinery configurations); and in the case of the Santos field (offshore São Paulo) failure to agree on a price with the risk contractor who found the gas (the first potentially commercial field discovered since the risk contracts were initiated). It is noteworthy that some of the private oil companies, whose marketing operations have been relatively unprofitable and who have been limited regarding other opportunities by the Petrobrás monopoly, have sought to get into the natural gas distribution business.

There should also be wide international interest in Brazil's experience in the commercialisation of alcohol fuels. Although for the present this programme is based on ethanol derived from biomass, some of the methods could be suitable for methanol derived from natural gas or other mineral sources and used for transportation. This option could be of interest to many more countries in the future, although Brazil's ethanol programme itself is somewhat vulnerable to declining oil prices. (Brazil's ethanol programme is discussed in a forthcoming ILO publication, *Ethanol, employment and development*, by A. Pereira.)

Many of the policy issues have had to do with conflicts among the various goals being pursued by Petrobrás. Its mission to find and produce expanding volumes of crude as rapidly and cheaply as possible tends to entail more recourse to foreign suppliers and contractors than can easily be justified under the other policy goal of stimulating the development of Brazilian expertise and capital goods suppliers (30 per cent of the country's output of capital goods is bought by Petrobrás). So priorities have had to be rearranged from time to time. In 1976 policy was changed to allow foreign firms in to explore for oil under risk contracts. The decision to push into deeper offshore waters has also increased, at least temporarily, the share of imported equipment, as Brazilian firms had not developed the relevant expertise for these operations. It has been suggested that the policy of accelerating the production schedule of new discoveries through the use of an early production system involves risks of damaging the reservoirs and thus sub-optimising the volumes ultimately recoverable from the deposits in question. There is also evidence that the pressure to proceed at a very rapid pace has involved a heavy price in terms of safety: accidents involving production platforms (the most dramatic being the fire that destroyed the Enchova platform), helicopters, boats and divers have been all too frequent and, it would appear, avoidable. There is a conflict between the policy of import substitution and that of keeping costs of production down.

Given its size and human resource base, Brazil has been able to create a comprehensive petroleum industry in a historically short period. The industrial spinoffs and the enhanced national autonomy have been substantial and clear achievements. Increasing oil and gas output (as well as other factors such as the important alcohol fuel programme) are beginning to ease the foreign exchange constraint on economic growth that energy import requirements have imposed in the past. These benefits have been attained at a price. Exclusion of foreign oil companies may have retarded the reserve and production increases that are now occurring. A cost, quality and safety penalty has probably been paid as the country has moved along the infant industry learning curve, although there can be debate about how heavy it has been. There has been a cost in human lives and suffering from the accelerated push to find and produce more oil offshore as rapidly as possible. The relation of these costs and benefits is one of the main policy issues in Brazil.

China

China is by most standards a major petroleum country. There is evidence that oil was first exploited there as early as 2,000 years ago, and natural gas as early as the thirteenth century. The modern industry dates from the discovery of the Dushanzi field in 1897. The country produced 114 million tons of oil in 1984, making it the world's sixth largest producer, and its recoverable oil reserves have been variously estimated at between 1.8 and 2.6 billion tons. In 1983 it is reported to have drilled over 4,000 wells, about one-quarter of which were exploration wells. In the order of 1.5 million people are estimated to work in the industry, if both upstream and downstream activities are included.

Since the establishment of the People's Republic, China's petroleum history can be divided into three periods:

1. Sino-Soviet collaboration (1949-60). Extensive geological and geophysical work was carried out with the help of Soviet specialists, resulting in the

discovery of some 600 undrilled structures, and the development of over 40 of these into commercial fields. The huge Daqing field, which still accounts for almost half the country's oil output, was discovered in 1959.

2. Independence and self-reliance (1960-78). Output grew rapidly during this period; production more than doubled during each five-year period between 1960 and 1975. Significant discoveries were made at Shengli and, offshore, at Dagang. However, exploration was somewhat neglected.
3. Collaboration with foreign companies (from 1978). Concerned anew about the exploration effort, the Government has since 1978 concluded contracts with a number of foreign firms or consortia first for offshore survey work and then, on a production-sharing basis, for offshore exploration. The results to date have been generally disappointing. By 1985 renewed emphasis was being given to onshore prospects, and in March of that year it was announced that foreign companies would be invited to participate in onshore oil exploration in an area of 1.8 million km^2, in the country's southern provinces.

Two features of the current, collaborative phase of policy are especially noteworthy. One is that the motivation for an accelerated petroleum effort seems to be based in large part on the need for export revenues. About one-fifth of oil output is currently being exported, and it accounts for about one-fifth of export revenues. Underlying internal demand is growing, although it is being suppressed, for example, by rationing. It may be expected to continue to grow as the economy expands and the new policy of permitting privately owned motor vehicles takes effect. Providing adequate production to meet internal and export needs is a challenge which the opening to collaboration with foreign firms was intended to help meet.

The second feature to be noted is the high priority being put on transfer of technology and preference for local supplies in the production-sharing contracts. These provisions are actively monitored through Joint Management Committees comprising representatives from the China National Offshore Oil Corporation (CNOOC) and of the foreign contractor. These committees review work programmes, procurement decisions, personnel and training questions and decisions about the commerciality of individual fields. The committees have been a potential source of friction between the CNOOC and foreign operators, but their existence is a manifestation of the Chinese Government's determination not only to accelerate its petroleum effort but to capture for its own long-term benefit as much as possible of the expertise and spin-off effects of the offshore activity.

Current production is concentrated mainly in the north-eastern parts of the country, which were already among the more industrialised parts. The development of the Daqing field in Heilongjiang Province with its associated pipelines and other infrastructure has reinforced this tendency. The offshore effort has been concentrated in the Bay of Bohai, which is sufficiently close to those regions to tie in to the same supply networks, and in the South China Sea. The latter effort has involved the development of new supply bases in the south, but their fate depends on the outcome of the exploration there, which has so far been modest. There is some speculation that the major discoveries of the future may be made in the remote north-western part of the country, but exploration there is hampered by the absence of roads, railways, communications and other basic infrastructure. It is probable too that the most important onshore reserves of natural gas are located in the far west, far removed from their logical markets and not presently served by the pipelines needed to transport them. Present natural gas output is modest (12.1 billion m^3 in 1983), but estimates of reserves vary enormously, and the country has been even less thoroughly explored for natural gas than it has for oil – by one estimate only 10 per cent of the sedimentary basins have been thoroughly explored (Kambara, 1984). Perhaps the most significant discovery to date of the offshore effort is ARCO's natural gas find south of Hainan Island; its reserves have been estimated at 90 billion m^3. Development is to be completed by 1992, at which time 3.25 billion m^3 per year of gas is to be produced and distributed via over 1,000 km of pipeline to various cities on the mainland and used for electricity generation and as industrial and domestic fuel.

Côte d'Ivoire

Offshore exploration began in the Côte d'Ivoire in 1970. The country's first discoveries were made by Esso in 1974 (Belier) and by Phillips in 1979 (Espoir). At the end of 1984 total oil production amounted to 22,000 b/d, roughly equal to the volume required for internal consumption. There is some associated gas at both Belier and Espoir, and Phillips has also found a more substantial non-associated gas deposit in a block adjacent to Espoir. In the early 1980s there were hopes that Espoir would turn out to be a major producer, and articles appeared in the press which foresaw production of as much as 200,000 b/d or even 500,000 b/d by the mid-1980s. In the event, such a bonanza did not materialise, and the outlook in 1985 is much more modest, by a factor of ten or so. Exploration activity has declined, as the relatively deep water, expensive operating conditions and worsened oil price outlook render even less economically attractive geology which seems to be judged of only limited potential.

During the few years between the discovery of Espoir (which coincided with the second round of international oil price escalation) and the clarification of its actual potential, public expectations showed some signs of soaring in typical oil boom fashion. With some lapses, the political leadership generally downplayed, in its rhetoric and investment decisions, the petroleum discoveries and their economic implications for the country. Stress was put on the country's agricultural and agro-industrial vocation; agriculture accounted for about 80 per cent of the country's employment and the lion's share of its foreign exchange earnings. Relatively little commitment was made to large prestige projects whose realisation or justification depended on sizeable petroleum revenues. Those projects which could be so characterised owed their existence more to the euphoria generated by the booming agricultural commodity prices of the mid- to late 1970s than to expectations of petroleum wealth, although that may have contributed. This prudence and maintenance of traditional priorities was vindicated when the petroleum market turned down and the country's petroleum potential came to be seen in more modest terms.

As of mid-1985 it appeared that the country might have more hydrocarbon reserves in the form of gas than of oil. Given the weakness of gas export markets – even for larger accumulations than those of the Côte d'Ivoire – the exploitation of this resource depends on finding local uses for it. A project has been under discussion for some time between Phillips and the Government; it may have been agreed by the time this report is published. It would involve the provision of gas to several users concentrated in the industrial area of Vridi, near Abidjan. These comprise an electric generating plant, some light industry, and (as fuel) the SIR (Société ivoirienne de raffinage) oil refinery. Fuel oil and other petroleum products would be displaced, thus improving both the net petroleum trading position and the financial viability of the refinery, whose losses have been a heavy drain on the state budget for several years. Employment generation would *not* be one of the major benefits, at least not in direct terms nor if compared with the alternative possibility of building another hydroelectric dam at Soubré. In a carefully organised, 31-month training programme designed by Phillips to prepare the staff needed for the gas operation, numbers foreseen are a meagre 22 recruited and trained, 11 hired.

The main effects, experienced or foreseen, of the Côte d'Ivoire's petroleum development, are the macro-economic benefits it has brought and the associated human resources story, which is related in Chapter 5. There have been few industrial spinoffs, as goods and services required are almost entirely imported, and little downstream activity – leaving aside the rather unfortunate example of SIR – has so far developed. Local effects, whether positive or negative, are neither sizeable nor visible, given that production takes place offshore and that the onshore activities are located near traditional port and industrial areas. It is worth noting that the continuity of leadership may have contributed to the considerable progress the country has achieved in the petroleum sector in a short period. Mr. Paul Gui Dibo has headed the Ministry of Mines and the State Secretariat which preceded it since the early 1970s. Many countries suffer from excessively rapid turnover in key petroleum positions; few have experienced such a long tenure.

Ecuador

Although oil has been produced in Ecuador since 1925, the country only became a significant producer, an exporter, and a member of OPEC in the early 1970s. That is when production came onstream from a discovery made in 1967 in the eastern, Amazonian region of the country by a Texaco-Gulf consortium. As this boom in production very nearly coincided with the initial phase of rapidly rising prices, the economic effects on the country were magnified. Petroleum exports increased from US$834,000 in 1970 to US$1.4 billion in 1980, by which time they represented 56 per cent of total export revenues.

The volume of production was on a plateau for the decade from 1973, roughly in the range 180,000-220,000 b/d. Internal consumption rose rapidly, stimulated by a pricing policy, enshrined in legislation, which tied product prices to the costs of production. Exploration languished during this decade. The national oil company, CEPE, was established in 1972. The following year it began buying into the Texaco-Gulf interests. For some years thereafter there was more emphasis on developing CEPE as an institution than on stimulating new exploration. As rising domestic consumption began eating into the available export volumes, and the reserves/production ratio declined, the policy was shifted. In 1982 legislation was passed providing for risk contracts. The application of the legislation was progressively eased, especially under the Government of Febres Cordero, as a higher priority was given to attracting foreign exploration. In 1984-85 new contracts were signed with Occidental, Esso-Hispanoil, Belco, BP, Conoco and Texaco.

The 1970s were for Ecuador the time of the petroleum boom. With petroleum revenues rising rapidly, government expenditure, the relative share of the public sector in GDP, and income per head all grew steadily. Although the distribution of the windfall income gains was highly uneven, there was during the decade a marked improvement in several of the social indicators, suggesting that many of the benefits of expanded government expenditures based on increased oil revenues were reaching a substantial part of the population. For example, the infant mortality rate dropped by a quarter, the illiteracy rate was almost halved, the number of doctors per 10,000 inhabitants tripled, the rate of secondary school enrollment doubled; for the last two of these indicators, the improvement surpassed by far that in other Latin American countries. Some of the benefits trickled down to the poorest rural stratum, the Sierra peasantry, whose real incomes showed a modest increase between 1968-75 (Commander and Peek, 1983). Other parts of the energy sector benefited from the petroleum revenues, which were used in part to finance the development of the country's hydropower resources.

Remarkably, the injection of substantial additional income into the economy did not result for some time in an accelerated rate of inflation. With the exception of 1974, when consumer prices rose by 23 per cent, the rate in the remainder of the decade ranged from 10-14 per cent, only slightly above the 8-10 per cent typical of the years immediately before the oil boom. One important factor contributing to this result was the moderate rate of increase of wages and salaries. Another was the relative abundance of foreign exchange, which allowed much of the additional demand to be met from imports. The balance of payments, in contrast to this good record of cost containment, was in deficit on current account as early as 1976, and thereafter the volume of external borrowing and the weight of the debt service increased rapidly.

The years 1981-84 were difficult, as the delayed effects of some of the heavy spending and borrowing of the 1970s began to be felt at the same time as the oil market turned down.

Ecuador's oil activities (there is very little exploitation of natural gas) are mainly concentrated in the jungle areas of Amazonia and in the coastal areas of Santa Elena and Esmeraldas. Production comes predominantly from the Amazonian province of Napo, which at the time of the first oil discoveries had a population of only 24,000; by 1982 it had increased to 113,000. The infrastructure and the non-petroleum economic base of the province are scanty, and much of the indigenous population is Indian. As oil development took place in a relatively unplanned manner, a number of social problems have resulted. Underlying many of these was the lack of integration of the petroleum activities into the local scene.

The problems included inadequate physical and social infrastructure, drugs, prostitution and tensions between the relatively well-off petroleum workers and the rest of the population. In March 1984 these tensions culminated in protest actions in which broad sections of the population of Napo Province participated. Access points to the oilfields were blocked, leading to a production shutdown. The TransEcuadorean pipeline was dynamited. An attack was attempted on the gas processing plant at Shushufindi. In the refining centre of Esmeraldas workers joined the action. The common theme was the inadequate participation of the principal oil producing and processing areas in the benefits of that activity. The action lasted a month. The outcome was legislation creating new funds for the development of infrastructure in the two provinces involved, half at the disposition of the municipal authorities and half at that of the provincial level. Some 50 million sucres (about US$425,000) were available under these provisions in 1984. That amount was expected to increase about threefold in 1985, but the monetary authorities later changed the basis on which the allocation is computed, in such a way as to eliminate or reduce that expected increase. In addition to these government funds, the state oil company CEPE reacted to the events of March 1984 by creating a Fund for Local Development to support projects (principally infrastructural and cultural) in the areas where its activities are concentrated. About 128 million sucres were allocated to this fund in each of the years 1984 and 1985.

Egypt

Some oil has been produced in Egypt since 1911, but the country only became a significant producer in the 1970s. Output more than doubled between 1974 and 1976, and again between 1976 and 1984: by the latter date it had reached 870,000 b/d. Exploration and production were mainly conducted in the earlier part of this period by foreign companies operating under concessions, but in comparison with other countries in the region the State made its influence felt at a relatively early stage. As early as 1938 the main concessionaire, Anglo-Egyptian Oil Fields (Shell), was required to give preference to Egyptians in hiring (90 per cent of workers and 50 per cent of professionals had to be Egyptian). In 1948 a law was passed requiring a minimum of 51 per cent Egyptian equity participation in the oil companies operating in Egypt. Although not immediately or uniformly applied, this law set the tone for most of the next quarter of a century. The state enterprise (now called the Egyptian General Petroleum Corporation (EGPC)) was created in 1956, and concession-holders BP and Shell were nationalised in 1964. The policy in the 1960s favoured the creation of joint ventures between EGPC and the private oil companies, but only a few such contracts were concluded. By contrast, after 1973 the shift to production-sharing contracts – in combination with Sadat's "Open Door" policy and rising oil prices – led to a dramatic increase in exploration activity. About 130 production-sharing contracts were signed between 1973-83 (although as many as 30 were cancelled in 1983-84 as the oil market weakened). In most of these the EGPC did not play an operational role, as under the joint venture arrangements the operator was instead the foreign partner.

The shift to and vigorous execution of the more liberal production-sharing arrangements was motivated by the need to find reserves to sustain the Egyptian economy's growing requirement for petroleum fuels, for internal consumption and for export earnings. The reserves/production ratio fell from over 30 in the mid-1970s to around ten at the end of 1984. Stimulated by a pricing policy which kept most product prices well below international market levels, internal consumption has grown at double-digit annual rates for a decade or more; it practically tripled between 1973 and 1983. In recent years this growing internal requirement has perceptibly reduced the portion of petroleum output available for export (from 60 per cent in 1978 to 51.5 per cent in 1982/83). Egypt has come to depend heavily on this source of foreign exchange. Nearly 70 per cent of visible export earnings in 1984 came from petroleum. This commodity's role in Egypt's external accounts is even more dominant if one takes into consideration the remittances from Egyptians working abroad, mainly in the oil countries of the Gulf (an estimated US$3.3 billion in 1983), and fees derived from oil tanker traffic through the Suez Canal and from oil pipeline transport. The petroleum sector accounts for between 15-20 per cent of GDP, and it contributes in gross terms around one-third of government revenue. In light of these various indicators of dependency, the country has a clear interest in preventing its exportable surplus from being eroded by too rapid growth in non-productive forms of internal consumption.

One way of achieving this objective is by encouraging the substitution of natural gas for oil in internal markets. Efforts have been made in recent years to encourage both the utilisation of known reserves of gas and the exploration for new reserves. Gathering and separation systems have been put in place for associated gas that was formerly flared from the Suez fields, and the three major non-associated gas fields – Abu Madi in the Nile Delta, Abu Qir in the Mediterranean and Abu Gharadiq in the Western Desert – have also been put into production. Deposits are being reserved for local use ("National Gas Reserve") until 12 trillion cubic feet have been proved. Increments above that figure will be available for export. In 1984 the official figure was 7 trillion and rising, although spokesmen for some of the companies involved have used figures as high as 9 or even 12 trillion. Existing exploration contracts were amended after 1980 to provide an incentive to develop natural gas resources found while seeking oil, and these amendments have subsequently been incorporated in new exploration contracts. Stimulated by these measures, gas production from its commencement in 1975 has risen steadily to the level of 3.1 billion cubic meters in 1983. To date, utilisation has been primarily in industry – as fuel for cement and steel plants, and so on, and as feedstock for fertiliser plants. Power generation will, however, ultimately take a larger share, as gas is substituted for fuel oil in electricity plants. Household use still accounts for only a small volume, but is growing rapidly under the stimulus of very cheap prices. Total gas consumption increased in 1983 by nearly 50 per cent.

The oil and gas industry in Egypt employed in 1983 about 43,000 people, which represented some 0.35 per cent of the country's total employment. Less than 20 per cent of that number worked in exploration and production activities. Of these, in the order of 80 per cent worked in companies of mixed ownership, of which the largest was the Gulf of Suez Petroleum Company (GUPCO), a joint venture between Amoco and the EGPC. The number of expatriates among upstream petroleum employees appears to have varied in the range of 3-6 per cent over the past 30 years or so. The number dipped to the low end of that range in the 1960s, when policy regarding foreign involvement was rather strict, and increased in the 1970s with the proliferation of production-sharing agreements; in this period the expatriate role has been concentrated overwhelmingly in the upstream. As already seen, formal requirements were laid down for indigenisation in the petroleum industry as early as 1938. The general policy of favouring maximum use of Egyptian staff has continued, although it has not always been explicit in the laws and contracts. The success of the policy results in part from the major training effort dating at least from 1950, when the Anglo-Egyptian Oil Fields refinery established the first vocational training centre. Later the universities (e.g. Cairo and Ain Shams) and state institutions such as the Institute of Petroleum and Mining Engineering (Shebin El Kom) and the Egyptian Petroleum Research Institute took on the major training role for professionals, while skilled workers and technicians were trained mainly in the state petroleum enterprises. Altogether some 12,700 employees participated in the national petroleum companies' training programmes in 1983. Although technicians are sometimes in short supply, there is a surplus of trained professional people. One result has been the use of graduate engineers as technicians, jobs for which they were over-qualified. Another has been the migration of many petroleum sector specialists to the Gulf countries, especially during the boom years of the 1970s. It is not clear what role various factors have played in creating this oversupply, but it is interesting to note that in 1964 the Institute of National Planning made a massive overestimate of the future manpower requirements of the industry (80,000 employees were foreseen for 1985 in exploration, production and refining; the actual figure did not exceed 25,000). Such estimates must have played a role in stimulating what turned out to be the excessive growth of petroleum-related training programmes.

The industrial benefits (backward-linked) captured by Egypt as a result of its petroleum activity take mainly the form of services rather than manufactured goods. For construction – which accounts for an estimated 30 per cent of petroleum sector investments in Egypt – and for drilling, maintenance, transport (air, water, road) and engineering, a number of local firms have developed into important suppliers for the local industry. Most but not all of these have equity links to major international firms. Most equipment, however, continues to be imported, even for items, such as pipe, which are manufactured locally. In the forward-linked sectors the achievements are more notable. Refinery capacity of about 370,000 b/d (end-1983) is being significantly expanded (to keep pace with rapidly increasing consumption) and upgraded in part to cope with greater volumes of fuel oil being backed out from power generation by natural gas). Much of the design and construction work for this programme is being done by local companies such as ENPPI (Engineering for the Petroleum and Process Industries), Petroleum Projects Technical Consultation Company (both subsidiaries of the EGPC) and others, albeit on occasion in collaboration with Bechtel or other foreign enterprises. There has also been a petrochemical industry, aimed mainly at local markets, since 1950; it is making increasing use of natural gas feedstocks.

India

Two related themes have loomed large in India's petroleum policy debate since the country became independent. One is the foreign exchange burden represented by the country's requirement for petroleum imports. As early as 1954 petroleum (crude and products) accounted for 15 per cent of the total import bill, or about three-quarters of the food import requirement (Pachauri, 1977). After 1973 the oil import bill increased dramatically: oil imports amounted to over 30 per cent of the country's export earnings in 1975/76 and, after the second oil shock, to 78 per cent in 1980/81. The need to reduce these numbers by importing more cheaply and/or by finding and producing more oil and gas in India was one factor which, together with nationalist and ideological considerations, led the Government to transfer control of the industry from predominantly private and foreign companies to newly created national public enterprises. The extent to which India's petroleum activities should be conducted by state companies is the second major theme.

Downstream, the problem (in 1955 and again in 1960) was the Government's wish for the international companies to substitute low-priced Soviet crude for their own crude streams for their Indian refineries, and their reluctance to do so. Upstream, the Government wished for the companies to undertake more exploration in India in the 1950s and 1960s than they were inclined to do under prevailing conditions. As a result, the Government moved over a period of years to take control of the industry into Indian hands. In 1959 the Oil and Natural Gas Commission (ONGC) was formed and became the Government's principal instrument for exploration and production activities. In 1961 the Government took half the shares in Oil India Limited (OIL), the other half being held by Burmah Oil Company. This became the smaller of the two Indian upstream companies. Nationalisation of the upstream companies was completed in 1981, when the Government acquired the other half of OIL as well. Similarly, apart from two joint venture operations, the refining industry today is also entirely nationalised. This nationalisation process was gradual and spanned the 1970s. Crude oil supplies and technical assistance from the USSR, for both refining and exploration activities, were an important feature of this story.

More recently, the Government has shown renewed interest in attracting international companies into exploration activities. Blocks were submitted to international bidding in 1981, but the conditions were not regarded by the industry as favourable. Chevron was the only major company which signed a contract; it has since withdrawn.

India's reserves, production and self-sufficiency ratio have been rising at a rapid pace in recent years; by 1984 the country covered over 70 per cent of its petroleum needs from internal production. In the five years to 1983/84 oil production increased by an annual average of 17.5 per cent and gas production by an average 16.2 per cent. The major development behind these figures was the coming onstream of the Bombay High field, located some 150 km north-west of Bombay in the Arabian Sea. This field in 1983/84 accounted for 67 per cent of national oil output, and its associated gas (not counting the non-associated gas in the nearby South Bassein field) for 55 per cent of total natural gas production. It was discovered and developed by ONGC.

The sector is manned at all levels preponderantly by Indian nationals. In 1984 only 140 of the 43,435 persons employed by ONGC and OIL were expatriates or foreign experts. That number in fact represented a sharp increase from the 11 foreigners so employed in 1975. The increase was made necessary by the shift of emphasis into offshore exploration and production. This expertise is supported by the country's formal education establishment and by a considerable petroleum training programme. The ONGC is setting up a Training and Executive Development Institute in Dehradun which is planned to hold short training courses for 3,000 executives every year. In addition, three Staff Training Institutes are to be set up during the current five-year plan period; they are to provide training for some 6,000 workers and supervisors each year.

Efforts have been going on for some time towards indigenisation of oilfield equipment and services, although cost and availability considerations have led to continued reliance on foreign supplies in many cases. An estimated 40-50 per cent of equipment is now supplied from Indian sources. This includes offshore platforms and jack-up rigs, drillship, onshore rigs, and so on. As an expansion of upstream activity is now envisaged, the Government is encouraging the development of new Indian manufacturing capacity in these areas.

The natural gas sector is expanding. Supplies are expected to increase nearly fourfold between 1983 and 2000. In 1984 the Government created a new independent organisation, the Gas Authority of India Limited (GAIL) to be responsible for the distribution and sale of natural gas and natural gas liquids and for developing the necessary associated infrastructure. One of its responsibilities will be the Hazira-Bijaypur-Jagdishpur gas pipeline project, which is designed to pipe non-associated gas from the offshore South Bassein field some 1,700 km across the country, to supply feedstock to six fertiliser plants that are to be set up along the route. Another pipeline from the Bombay High and South Bassein offshore fields is also planned to supply offshore gas to the city of Bombay for household use, and favourable consideration is being given – in view of the continuing availability of flared gas – to greater use of gas for power generation.

Malaysia

Although oil has been produced since 1910 in the area now comprising Malaysia, it is the development of offshore resources in the 1960s and 1970s that has given the country's petroleum industry the shape it has today. These are located off the states of Sabah and Sarawak, which are in East Malaysia on the island of Borneo, and off the state of Trengganu, which is on the east coast of peninsular Malaysia. As it happens, these are all among the least developed and least populated parts of the country.

Since the first offshore production began in 1968, development has been very rapid. Malaysia's oil output was 4,000 b/d in 1968, 90,000 in 1973, 280,000 in 1980, and 450,000 in late 1984. Estimated reserves have also been going up; in early 1984 they stood at 3.0 billion barrels, which was 80 per cent higher than the estimate two years previously. Natural gas reserves are three times as great as oil reserves, in terms of energy equivalence. A LNG export project, based in Bintulu, Sarawak, went onstream in 1983; it is based on a contract to supply Japanese markets for a 20-year period.

In mid-1980, after five years of rapid expansion of output, the Government adopted a depletion policy, which led to production cutbacks of 3 per cent that year and 6 per cent in 1981. However, owing to the need for revenue, the ceiling was removed in 1982, and production rose sharply again. In 1985 output was voluntarily held at 420,000 b/d in sympathy with OPEC, but on account of the recession a record 510,000 was targeted for 1986. In 1984 petroleum accounted for almost one-third of federal revenue and 27 per cent of export earnings.

Petronas, the national oil company, was created in 1974 as the Government's main instrument for protecting national interests in petroleum activities. The decision to establish a state company was influenced in part by the country's experience as an exporter of other commodities, such as rubber and tin, the trade in which was dominated by foreign interests. It was also inspired by the developments in the oil market in the autumn of 1973, from which the lesson was drawn that the strategic character of the oil trade made the creation of a state oil company advisable. Petronas administers the production-sharing contracts that have been negotiated with foreign-owned operators (Exxon and Shell account for most of current output). It is also an operator in its own right and, through its subsidiary Carigali, is now involved in exploration as well as production. Relations between the state company and the foreign contractors have been relatively constructive and free of major frictions.

Petronas' objectives have included the enlargement of employment opportunities for Malaysians in the industry and, within that, the promotion of the interests of Malays and other indigenous peoples (*Bumiputras*, as distinct from Malaysians of Chinese and Indian origin) in accordance with the New Economic Policy introduced in 1970 (for 20 years). Great strides have been made in these directions. Since 1973 the ratio of expatriates to Malaysians among senior staff in the industry has been reversed: roughly, it has gone from 3 : 1 to 1 : 3, and that in a group whose numbers have been expanding rapidly. Although the reverse discrimination implied by the New Economic Policy has been controversial at times, there have been significant advances for the *Bumiputras*. Companies' performance in these respects must be reported and is closely monitored by the Government. Preference is also given to local suppliers and contractors, especially *Bumiputras* and those from the oil-producing regions, provided they meet industry standards.

Half of the royalties paid under the production-sharing contracts are paid to the state governments in the oil-producing regions. This has given these local authorities means to develop some of the physical and social infrastructure that intensive petroleum development has created a need for. Even so, there has been debate about whether these funds have been adequate and whether enough has been done to distribute the benefits of the country's petroleum patrimony in the areas where the production is concentrated.

Pakistan

Although its modest oil output rose rapidly in 1984-85, Pakistan's petroleum sector has been notable to date mainly on account of its history of natural gas exploitation. For many years oil production had been rising in step with consumption, so as to provide around 10 per cent of total requirements. On the strength of a number of discoveries and extensions of existing fields, output doubled in 1984 to about 28,000 b/d, and increased further to 42,000 b/d by October 1985. This carried it to about one-quarter of consumption. Drilling activity has been stepped up since 1981: an average of ten wells per year were drilled during the period 1977-80; in 1981-84 the figure was 30. Exploration drilling offshore got under way in 1985. Since the current five-year plan calls for the average number of exploration wells to rise to 50 over the period 1983-88, it is not unreasonable to expect further increases in the self-sufficiency ratio. The Minister of Petroleum has predicted that the country would be self-sufficient in oil by 1989.

Notwithstanding these encouraging developments for the future of oil, natural gas has thus far been the star performer in Pakistan's energy scene. Its share of national commercial energy consumption in 1983-84 was 35 per cent, all of it locally produced, as against 39 per cent for oil, 87 per cent of which portion was imported. This puts Pakistan near the top of the developing countries' league in terms of the share of natural gas in commercial energy consumption.

The increase in gas consumption followed the discovery of a large deposit at Sui in 1951 by Pakistan Petroleum Limited (a joint venture of Burmah Oil Company and the Pakistani Government). In 1983-84 three-quarters of the country's gas output still came from the Sui field. Initial distribution was limited to a small number of industrial users in Karachi, to which the field was connected by a 560 km pipeline built, together with a gathering and purification system, between 1953 and 1955. Over the years the network was expanded, in 1960 to Multan in the Punjab and later to Lahore and Islamabad. In 1977 a second line was added to the southern system. More recently Quetta, the capital of Baluchistan, has been tied in. Newly discovered fields have also been connected to the supply system. As the infrastructure expanded, new classes of users were added. As of 1983-84 the distribution of consumption by sector was as follows: fertiliser, 32.4 per cent; power, 24.7 per cent; general industry, 23.5 per cent; residential, 9.4 per cent; cement, 7.0 per cent; commercial, 3.0 per cent.

The rapid rate of growth of gas consumption (annual average 8.1 per cent 1983-84/1969-70) was stimulated by a pricing policy which deliberately kept gas cheaper than alternative fuels in order to substitute it for imported energy, especially oil. In later years the problems associated with this policy have been recognised – disincentive to exploration and to conservation, encouragement to energy-intensive industrialisation patterns, acceleration of inter-fuel substitution from non-commercial energy to gas, tendency for demand to outstrip supply. Consequently, gas price increases have been introduced at an accelerated pace since the early 1970s and especially since 1982; however, the price continues to be low by international standards. More remarkably, the substitution process has been reversed as several power plants and cement works were converted from natural gas to oil.

Almost the whole of Pakistan's gas industry is staffed with local personnel, as the country has a surplus of technical personnel in petroleum technology.

One of the problems associated with the rapid development of the gas production and distribution system has been an accident rate that has been regarded as unacceptably high. Government legislation and inspection and standards procedures have been adopted in an effort to remedy the problem. Some progress has resulted, more on the side of industrial utilisation than on the production side, but room for improvement remains.

Peru

The first oil wells in Peru were drilled in 1863, making it one of the world's oldest producing countries. For long it was a modest net exporter, but in the late 1940s and 1950 a pricing policy which encouraged consumption and limited rewards to producers led to growth in consumption that outstripped production; the country became a net importer in 1958. In the 1970s new reserves were found in the Amazonian jungle, of sufficient size to justify building a Trans-Andean pipeline, and in 1979 Peru again became a net exporter. As of 1984 about one-third of its total output of 180,000 b/d was exported.

No major new discoveries have been made since the mid-1970s, and proved reserves, which peaked in 1981 at 835 million barrels, have declined to about 645 million barrels. In the absence of new discoveries the export surplus will soon dwindle. Since petroleum production contributed about 18 per cent and 25 per cent, respectively, of foreign exchange earnings and central government revenues in 1983, this prospect is viewed with grave concern, especially against the background of the country's heavy external debt service obligations. In 1984-85 efforts were intensified to attract new foreign exploration contracts and to gain increased investment funds for the state oil company, Petroperú, for its exploration programme. One problem has been a tendency for the central government to appropriate a large share of the company's cash flow for other purposes rather than allowing it to be ploughed back into the search for petroleum. The attempt to attract new foreign investment was complicated, following the election of the García Government in 1985, by a dispute between the State and existing contractors – Occidental, Belco and Bridas – regarding past tax credits and related investment spending.

Petroperú was formed in 1969, following the nationalisation of International Petroleum Company (IPC) – an Exxon subsidiary – the previous year. Much of its staff, including technically skilled personnel, was inherited from IPC. During the period 1975-78 a policy of restraining salary increases in the company, at a time when there were attractive opportunities outside the company and the country, led to the loss of many experienced engineers and technicians to other employers. Although this weakened the company, the salary policy has subsequently been changed so as to allow competitive remuneration to be offered, and alternative opportunities have recently been more limited, so it has become easier to keep qualified staff. About 40 per cent of the 25,000 people employed in the Peruvian petroleum industry work for Petroperú.

Transport, refining and distribution of oil are the monopoly of Petroperú. Upstream operations are partly managed by Petroperú and partly by foreign contractors. One-third of reserves and production are in fields operated by Petroperú, and two-thirds are in those operated by private companies, mainly Occidental (in the jungle) and Belco (offshore) – although the latter's role was cast in doubt in 1985 by an investment/tax dispute with the García Government. Jungle production is twice as great as that of the coastal/offshore areas.

Since 1968 the technical capabilities of Peruvians working in the industry have increased significantly, so that at most stages of the exploration/development/production chain a larger part of the effort can be supplied by national means than was the case earlier (for details, see Chapter 5). For the more technologically demanding operations, however, there is still heavy reliance on foreign expertise. There has been only slight progress towards capturing sizeable supply contracts for petroleum activities for Peruvian industry. The main contribution has been in construction, transportation and basic industrial goods. One exception to this general point can be found in Talara, where a co-operative of engineers and workers – mainly welders – was fabricating platforms (five or six a year) for offshore development work, under Belco supervision.

The two localities mainly affected by petroleum operations are the coastal town of Talara and the Amazonian settlement of Iquitos. The former is very much an "oil town" that is the base for what may be the oldest continuously producing oil province in the world. Over 6,000 workers are directly employed by the oil companies there and another 10,000 in service companies or other associated or indirect activities. In the days of IPC and for some years thereafter, Talara was in effect a camp, where basic services were provided by the company, not by any municipal authority, and where a fence surrounded the oil installations including the houses of staff. In 1972 it was declared an "open town", not an oil camp, but the dualistic character of the settlement remains. Iquitos in 1984 was a town of about 100,000, located on a navigable tributary of the Amazon. It serves as the administrative centre and supply base for most of the oil activity in north-eastern Peru. It has the character of a base town and shares some of the problems of the producing regions in the Amazonian parts of Ecuador, already described.

In an effort to ameliorate these problems of impact on producing areas, companies in the extractive industries are obliged to pay a sum equivalent to 10 per cent of their net income to the regional authorities of the areas where they operate, in these cases to the Development Corporation of Piura, for Talara, and the Development Corporation of Loreto, for Iquitos. These contributions in the years 1980-84 amounted to about US$320 million for the two regions combined. This "Canon", as the assessment is called, is an imaginative attempt to address the problem of impacted producing areas. Problems have however been encountered in the distribution and use of the funds, in part because the way in which the funds are used depends on the distribution of political power at the regional level; this may not assure that the localities most affected by petroleum activity receive most of the funds.

Thailand

The discovery (by Union Oil) of a significant natural gas field in the Gulf of Thailand in 1972 and (by Shell) of a modest-sized but commercial onshore oilfield at Larn Krabue near the centre of the country in 1981 has turned Thailand into a petroleum producer and has substantially altered its energy situation. Gas production began in 1981, when the 475 km underwater pipeline – the world's longest – needed to bring it ashore was completed. Oil production from the Sirikit field at Larn Krabue began in 1983. In terms of current levels of output and of reserves, the natural gas deposits are far larger than those of oil (proved gas reserves in the Union Oil concession of about 700 million boe vs. estimated proved oil reserves of 45 million barrels). These figures exclude Exxon's onshore gas find in the north-east of the country and Texas Pacific's "B" gas field farther out in the Gulf of Thailand, both of which are so far undeveloped.

These petroleum resources became available at a time when Thailand's consumption of petroleum products was rising rapidly: the average annual increase from 1975-83 was 7.4 per cent. The implication is that in absolute terms the volume of petroleum imports has not fallen significantly – from 214,770 b/doe in 1981 to an estimated 212,500 b/doe in 1984. In trade terms, at least until the end of 1983, the situation worsened. From 1979 to 1983 the share of petroleum products in total commodity imports rose from 22 per cent to 25 per cent, and net petroleum imports as a proportion of total exports increased from 30 per cent to 40 per cent. This occurred as domestic oil output (including condensate/NGL) was rising from virtually zero to 25,000 b/d, and gas production was increasing from zero in 1979 to 56,731 million cubic feet in 1983 (approximately 26,000 b/doe). The figures for 1984 and 1985 will show further significant increases in gas production, but that will not alter the conclusion that the coming onstream of Thailand's petroleum resources has thus far displaced part of the growth in petroleum imports that would otherwise have occurred but has not significantly reduced the absolute level of those imports.

There are two notable features of Thailand's strategy for utilising its natural gas resources. One is the phased, two-part programme of market development. The other is the way in which it has been carefully integrated into the country's general industrial and regional planning.

In order to derive benefit as early as possible from the available gas supply, it was first provided to markets requiring the shortest lead time and simplest economics and investment planning: two electric power plants, in South Bangkok and Bangkapong, supplied from the same transmission line. Subsequently, a 179 km extension of that line was built to allow two cement factories to be supplied. The displacement from these uses of imported fuel oil by domestic natural gas accounts for much of the balance-of-payments gain from Thailand's petroleum development, a gain approximately equal in 1983 to one-third of the country's merchandise trade deficit. There has resulted a dramatic shift in the composition of petroleum imports. In 1981 fuel oil for power generation accounted for 23 per cent of petroleum imports; for 1985 the figure was expected to be about 4 per cent.

Meanwhile, planning was put in motion for longer lead-time, higher value-added projects to be supplied by future increments of gas output. These projects were integrated into a major development plan, the Eastern Seaboard Project, which was intended not only to optimise the economic benefits from the gas resource but also to give a significant new orientation to the pattern of land use planning in the country. The idea was to build a gas separation plant and related industries and port near the landfall point of the natural gas pipeline. All this development was intended to form an alternative growth pole to Bangkok, thus easing some of the population and other pressures on the capital.

The gas separation plant which is the centre-piece of this project came onstream in late 1984. It will ultimately produce about 450,000 tonnes of LPG, as well as continuing to provide natural gasoline, ethane and methane for power generation. (The consumption of LPG has been rising rapidly in Thailand and in 1985 was approximately equal to that 450,000 tonnes figure. Most of this demand growth came from transportation – whose use of LPG increased twelvefold from 1980-83 – as Bangkok taxis and others switched to a less polluting form of fuel.) The intention was to use the ethane and the propane fraction of the LPG as feedstock for a series of petrochemical plants to be built and commissioned as available capital and economics confirm their viability. These include an olefin plant and plants to produce high density polyethylene, polypropylene, vinyl chloride monomer, methanol and nitrogenous fertilisers. A second gas separation plant and a LNG liquefaction terminal could be added at a later date if appropriate in the light of developments in the meantime.

These projects are capital-intensive, but they none the less would create significant numbers of jobs, and they are jobs, moreover, with high value-added. If fully executed, this plan is expected to create by the year 2000 between 9,000-10,000 jobs directly and another 6,000 or so indirectly. For comparison, the number employed in manufacturing in Bangkok in 1981 was 175,000. Hence the relative concentration of new employment on the Eastern Seaboard, centred at Map Ta Phut, about 175 km south-east of Bangkok, would be substantial. An additional 200,000 urban residents would be expected in the area, over and above what would be expected in the absence of this development, bringing the population of the region to 2.1 million.

Planning for this development has been remarkably thorough. Following the formulation of strategic guide-lines for the project by an inter-ministerial committee, a detailed study was commissioned from an independent consultant, proposing urbanisation and land use arrangements, including the necessary physical and social infrastructure and the phasing of development.

All of this has been done in an attempt to derive maximum benefit from Thailand's newly discovered gas resource, without which the project would not have been conceivable. It is sobering to reflect that, despite the meticulousness of the planning, changes could occur – and by the end of 1985 were occurring – in the international oil and gas markets, over which Thailand has little control, which could jeopardise the economic basis for some of the industrial projects on which the development depends. That is the nature of the industry.

Major exporters

Although not the focus of this study, for reasons already explained, countries which export petroleum in volumes that are large enough to have powerful effects on their macro-economic performance present characteristics which are worth noting. Three cases are described here. Two are populous countries, one of which (Nigeria) is a member of OPEC. The third is a small island country with a population that is significant relative to its area (Trinidad and Tobago) and a volume of oil exports that is large relative to the size of its economy.

Mexico

The oil boom of the 1970s was not the first in Mexico's experience. A half-century previously, the country had been one of the world's leading producers; in 1920, for example, with an output of over half a million b/d it ranked second after the United States. This earlier episode was marked by growing friction between the authorities and the foreign companies which then dominated the sector in Mexico, leading to the nationalisation of the industry and the creation of the state oil monopoly, Petróleos Mexicanos (PEMEX), in 1938. This took place in an atmosphere of bitterness and recrimination far different from the planned handover that took place during the nationalisation process, for example, in Venezuela in the 1970s. Forced to rely on its own resources, Mexico gradually worked its way along the learning curve. As it did so, its internal consumption overtook its production, so from the 1950s through the early 1970s the country was a net oil importer. Reacting to that situation, and to the 1973 price increases, an accelerated programme of exploration was embarked upon that resulted in the major reserve and production increases of the late 1970s. Oil output and exports, respectively, rose from 897,000 b/d and 94,400 b/d in 1976 to 2.7 million b/d and 1.5 million b/d in 1983. Official figures for proven reserves (which have been widely challenged) increased from 11.2 billion barrels in 1976 to 72 billion barrels in 1982 (Mendoza Berrueto, 1984).

The revenues generated by this growth of (higher-priced) oil output helped the López Portillo Government to recover from the difficult financial situation that it had inherited from the Echeverría administration, to restore the balance of payments to health and to resume, and surpass, the high rates of economic growth that the country had enjoyed from the 1940s to the early 1970s. Since the overwhelming majority of workers, including all the employees of Pemex, were Mexican, the employment generated by this activity was predominantly captured by the national economy (of the 180,000 people occupied in petroleum activities in 1983, roughly half were full-time employees of Pemex and its affiliates (Norconsult, 1983b). Moreover, a substantial share of the capital goods and other supplies required by the industry (including most pipe, valves, pumps and shallow drilling equipment) are supplied from national sources, which in the years since 1938 and especially since the establishment of the Mexican Petroleum Institute in the 1960s have acquired the capability to cater to these needs of Pemex. Even when expanded demand reduced the share of orders which they were able to meet, Mexican suppliers still provided around 60 per cent of the investment goods required by

Pemex (Sterner, 1982). Petroleum revenues have also been seen as resources for furthering development in other ways, including support for transport, infrastructure and for the Mexican Food System (SAM), aimed at increasing agricultural production (Sanders, 1982).

There was also a negative side to this rapid expansion of output, and it has been well publicised. At the macro-economic level, items include an increase in the rate of inflation (15 per cent in 1976 to 70 per cent in 1981), the problem of external indebtedness (Pemex's share of which was an estimated 30 per cent), and an overvaluation of the peso in the late 1970s which penalised Mexican-produced tradeable goods, both agricultural and industrial. There were problems of social and regional imbalance. The former relates to problems of maldistribution of income, including the difference in rates of remuneration between petroleum workers and others. The regional problems pertain to the impact of petroleum development activities on the sparsely populated and predominantly agricultural areas near the new oil fields, for example in the state of Tabasco and especially in the city of Villahermosa. Rapid and relatively unplanned growth in these areas resulted in shortages of housing, related infrastructure and urban services in general; an acceleration of inflationary pressures; disruptions to the local labour market caused by the inflow of skilled workers; environmental nuisances; and a trend towards a dualistic society (Allub and Michel, 1982).

The remarkable thing about this pattern of oil boom effects is that it occurred despite considerable national awareness, growing out of the country's experience earlier this century, of the dangers of an "oil economy", or of what was sometimes referred to as "Venezuelization". "At no point did it occur to energy policy-makers of the 1970-75 period [when exploration activity was accelerated] that Mexico should again become (as it was between 1900 and 1930) an export-oriented 'oil country', another Venezuela" (Wionczek, 1982). Yet the population and other pressures were such that expansion occurred until, by 1983, it was proudly stated that the petroleum sector accounted for "10.2 per cent of the nation's GDP ... 78.3 per cent of total merchandise exports ... 36.8 per cent of total tax income and ... more than one-third of public investment" (Mendoza Berrueto, 1984). The pain of adjusting to declining oil revenues in the mid-1980s was therefore all the greater.

Nigeria

Nigeria's petroleum experience, in relation to its development, has been rather better documented than that of most countries (even for general Western television audiences, as shown by the screening in the United Kingdom of a film on the subject in the spring of 1984). Despite its special features, especially the fact that the oil boom followed soon after the end of a devastating civil war, it is appropriate to indicate here what messages that story conveys that are pertinent to this study. One of the main themes concerns the way the Government spent the much enlarged revenues provided by the oil boom of the 1970s, and the consequences of those choices. Another related one has to do with the extraordinary effort devoted to human capital formation, both in the petroleum industry and more generally.

The discussion which follows examines the nature of the development which took place on the basis of Nigeria's petroleum resources. It is of course basic that, whatever shortcomings may have characterised the use to which the resources were put, in direct terms they contributed significantly to the country's GDP, government revenues, foreign exchange earnings, energy supplies and – if indirect effects are taken account of – employment generation.

A notable feature of the expenditure pattern – in contrast, for example, to that of Trinidad and Tobago – is the speed with which the new revenues were spent. Government expenditures in 1975-76 were more than six times the level of three years earlier (Bienen, 1983). Given the scale of need in the country and the magnitude of the new resources, the Government chose to move quickly in order to be seen to be using available means to address recognised problems. Another consideration was the need, following the civil war, to rebuild the country and strengthen the central Government by developing the physical infrastructure needed to tie the regions together. In combination, these factors contributed to a pattern of expenditures in the mid-1970s in which public sector wage increases and large investments in education and transportation were major elements. Much of the expenditure was channelled through the states, which became dependent on the federal Government for revenue. This political purpose of cementing ties between the federal level and the states, in the aftermath of the civil war, played a major role in influencing expenditure decisions. The urban areas, where expenditures could be effected quickly through consumption subsidies and construction of housing for low income groups, benefited more than rural areas. Income distribution appears to have become more unequal as between rural and urban dwellers and probably also on an inter-personal level (Bienen, 1983; Collier, 1983).

The effects of these policies that have been widely noted are a sharp acceleration of inflation and a decline in the agricultural sector. The two were linked, inasmuch as a contributing factor to inflation was the relatively inelastic response of the agricultural sector to the increased demand for food brought about first by the end of the civil war and then by the growth in urban demand stimulated by the new oil revenues. There was a dramatic increase in money supply in the mid-1970s, as the Government monetised the increased oil revenues (Madujibeya, 1976). Relief of the demand-supply imbalances through imports was limited by inadequate port and transportation facilities. Inflation increased from 6 per cent in 1973 to a peak of 33 per cent in 1975, and then declined steadily through 1980. The apparent acceleration in the decline of agricultural output that occurred in the mid-1970s was especially marked in the export sector, which was hit by a changing pattern of consumption (higher import propensity) associated with urbanisation and by the appreciation in the real exchange rate brought about by the increased oil income.

On the positive side, the emphasis that the Government put upon the training of Nigerians to take positions at all levels of responsibility in the petroleum industry bore fruit relatively rapidly. Encouraged by a 1969 legal provision requiring that all skilled, semi-skilled and unskilled workers be citizens of Nigeria and that 75 per cent of the persons in managerial, professional and supervisory grades should be Nigerians within ten years from the grant of an oil mining lease, the companies have made a considerable effort through training programmes and scholarships to comply. The Nigerian authorities have also made a substantial allocation of resources and a commitment to institution-building in support of this goal. By 1979 the companies engaged in exploration and production activities had, according to one account, 50 Nigerians and 44 expatriates at senior management level and 1,255 Nigerians to 295 expatriates at professional level (Norconsult, 1983a). One of the larger operating companies reduced the number of expatriates in its operations from more than 800 at the end of the civil war to about 400 in 1977 and approximately 150 – many of whom reported to Nigerian supervisors – in 1985. This has not been accomplished without friction between, for example, young Nigerians with superior formal educations and their less well-educated but more experienced expatriate colleagues. And there has been controversy about the extent to which these results have been achieved by "window dressing" (for example, by creating redundant or duplicate posts for Nigerians, or by shifting expatriates to the payroll of subcontractors, who do not have the same training obligations). But these are questions of degree, and they do not leave in doubt the large measure of solid success that has been achieved in transferring the management of Nigeria's petroleum sector into Nigerian hands.

There is less achievement to report in the area of industrial spinoffs of petroleum activities in the country. On the supplier side, given the state of industrialisation of the country – the absence, for example, of an indigenous steel industry – relatively little has been locally provided. An estimate made in the mid-1970s put at only 15-20 per cent the share of the petroleum industry's expenditure that went for locally produced goods and services, and this consisted mainly of such things as electricity supply, payments to local contractors, telephone and postal charges, local rents, compensation for farm land, customs duties, etc. (Madujibeya, 1976). In terms of forward linkage, there has been much talk of and planning for energy intensive industrial schemes (steel complex, LNG), but they have been slow to be realised and time has not been favourable to their economic justification.

The main effort has been in education, not only in the petroleum sector but more generally, and in infrastructure. Expenditure on education increased from 13 million Naira in 1973/74 to 601 million Naira, the largest item in the recurrent budget, in 1976/77 (Bienen, 1983). A programme for making available universal free primary education was a major element in this effort. Capital expenditures on transport and communications increased from 154 million Naira in 1973/4 to 1.8 billion Naira in 1976/77, and stayed at a high level through 1980; much of this went for road building. Agriculture received little, except in so far as it benefited from the improved infrastructure. Given the haste with which these funds were spent and the opportunities for corruption which existed, there is room for question about how efficiently these sectoral allocations were carried out. There were also questions about the extent to which the new programmes were sustained in the leaner years of the 1980s. But the main point is that time will be required to judge how well advised the allocations themselves were. "How successful the Nigerian strategy of transforming oil wealth into human infrastructure capital (hence into non-traded assets with a wide developmental impact) has been, will only become apparent after a number of years, required for effects on productivity to be demonstrated" (Bienen, 1983, p. 46).

Trinidad and Tobago

The island of Trinidad, located 11 km off the Venezuelan coast, has since the early years of this century produced oil in steadily growing amounts, until its output peaked in 1978 at 11.6 million tonnes (230,000 b/d), falling subsequently to 9.0 million in 1984. It has sizeable and still largely untapped reserves of natural gas (published proved reserves end-1983 of 13.1 trillion cubic feet, equivalent to 330 million tonnes oil equivalent, or about four times the equivalent of recoverable oil). The country's population is 1.1 million.

Trinidad and Tobago is very much a petroleum economy, in the sense that oil and gas account for a large share of GDP, government revenue and export earnings (37 per cent, 61 per cent and 87 per cent, respectively, in 1981). Since its oil reserve base is being depleted, and its export refining sector has become vulnerable because of world-wide excess capacity, it presents in especially acute form the challenge that faces most petroleum exporters: how to convert the petroleum resource within the available production timespan into an economic base that can be sustained without petroleum. In Trinidad and Tobago this challenge has not been rendered easier by the boom in private and public consumption – including heavy food and fuel subsidies – that was fuelled by the petroleum price rises of the past decade, nor by a pattern of high wage costs and an industrial relations scene in which wage restraint was not easily exercised.

In the mid-1970s the Government decided that a key part of its strategy for coping with this challenge would be to develop a series of natural gas-based industrial projects at the site of Point Lisas on the west coast of Trinidad. These included ammonia plants, an integrated steel-making complex, a gas-fired power station, and methanol, urea and cement plants. Further schemes under consideration have included a joint methanol and methyl tertiary butyl ether (MTBE) plant and a LNG export project. Since the time when this strategy was adopted, international market developments have affected the economics of these projects adversely, leading in some cases (e.g. LNG, MTBE) to their postponement or cancellation and in others (ammonia, methanol) to a reduced or negative rate of return. The steel company lost TT$190 million in 1985.

Financing for these projects has been facilitated by various special funds for long-term development which were created after the jump in petroleum revenue in 1974 for the purpose of immobilising some of the windfall gains until the development programme was ripe to absorb them effectively. Over the period 1974-80 over half of the central Government's revenue from petroleum was appropriated to these funds, representing over TT$7 billion in all. The balances are held on deposit in the central bank and earn interest which accrues to each individual fund (there are over 40 funds in all) until spent. Until 1979 annual expenditures were consistently below new appropriations, so the balances continued to grow. A high proportion of expenditure from the funds represents public investment in the various development projects and associated infrastructure.

Another key element in any long-term development strategy is investment in human capital. Most foreign petroleum contractors are required both to give preference to local candidates over expatriates and to engage in training activities "with a view to training nationals … to replace foreign personnel as soon as reasonably practicable and to affording nationals … every possible opportunity for occupying senior positions. …" (Petrol. Reg., 1970, s. 42*g*, cited in Khan, 1984). The relative number of expatriates was quite low as early as the late 1960s (about 2 per cent of all personnel), and it has stayed about the same or declined since then (Turner, 1982). Trinidadian efforts to support this goal through national training efforts have been slow to get going. University-level courses in petroleum engineering were offered in the country for the first time in 1977. Most reliance has been put on the contractors. One of the companies operating there, Trinidad-Tesoro, has endowed a chair in petroleum engineering at the University of the West Indies. It and other companies also operate their own staff training programmes and also send some nationals abroad on scholarships. Despite these efforts, shortages of skilled personnel needed for the execution of the industrial development programme persist, combined with an unemployment problem among the unskilled.

Chapter 4

Macro-economic issues

In all the countries examined in this study it is a widely accepted goal that for the foreseeable future oil and gas production should be maintained at present levels or increased (leaving aside tactical debates about the advisability of temporarily lowering production as part of an international effort to shore up prices). In many if not most countries that are not now petroleum producers, it is government policy to encourage the search for and, if found, the production of hydrocarbons. The reason for this nearly universal commitment to the goal of petroleum production is not obscure. It is the macro-economic benefits that are presumed to accompany petroleum production. The principal such benefits are these: a direct stimulus to economic growth through the introduction of a significant new natural resource input; an improvement in the foreign trade balance, through added net petroleum exports or reduced net imports, resulting in an indirect growth stimulus through the easing or removal of the balance-of-payments constraint on imported inputs; and an increase in government revenues, through the capture by the government of part of the rent incorporated in crude petroleum prices, as well as through taxation on petroleum products; at government discretion, these revenues may be invested or spent on current consumption, with or without an intended or actual redistributional effect. Other things being equal, these developments should lead to rising real incomes and hence higher standards of living for a significant share of the population.

It is appropriate to discuss these macro-economic effects here, for two reasons. First, most of this study deals with effects that are secondary or micro-economic, which have been less frequently examined than the primary or macro-economic ones. In suggesting that the former deserve more attention than they have tended to get, we should not lose sight of the fact that the macro-economic factors are much the more powerful, both in their real effects and in the weight which they are given by government policy-makers. That is true even for effects felt at the local level. The local impacts of petroleum activities at the worksite involve a geographical area that is only a minute share of the national territory, whereas those that are transmitted through the fiscal and monetary mechanisms touch small as well as large localities, farms and enterprises throughout the country. Second, the macro-economic effects are indeed positive, as indicated in the previous paragraph, in a *static* sense. The preceding chapter is replete with examples of government budgets and merchandise trade accounts in which the contribution of petroleum is substantial and, for that reason, is cherished; it begins to appear, moreover, immediately petroleum production is initiated. However, if viewed in *dynamic* terms, these effects are more complex and not necessarily so uniformly beneficial (see Seers, 1978; Gelb, 1981).

The nature of the policy questions involved has been effectively summarised:

> The dynamic issues facing the oil economy reflect policy options at a number of levels. Considering an individual country as an international price taker, and abstracting from institutional constraints on the level of production, the prime decision relates to the petroleum extraction profile over time. Next comes the split of income into consumption and accumulation, and the division of investment into real domestic capital formation and the accumulation of claims on foreign countries. All the above choices may be summarised as fundamental saving and portfolio decisions. Next come the policy issues relating to the appropriate sectoral distribution of domestic current and capital spending out of oil revenues. The subsequent level of choice concerns policy and institutional development to shape appropriate distributional, consumption, saving, investment and production patterns in the non-oil economy. The non-oil sectors will be subject to two sources of stress: changes in the intensity of demand and its sectoral pattern, and shifts in relative prices. The more rapidly it is desired to boost domestic absorption and the less flexible is the non-oil economy, the more violent will such stresses become (Gelb, 1981, p. 4).

It is not feasible in this study – and perhaps not at all, given the data problems – to analyse satisfactorily the ways oil producers have chosen to spend and invest their petroleum revenues and the effects that have flowed from those choices. But it may be useful to identify some of the typical effects and explain how they operate.

One problem is that the improved foreign trade position is likely to be accompanied by a strengthening of the real exchange rate, and this will have differential effects on various sectors of the economy. Tradeable goods and services, both those sold domestically and those exported, will become less competitive against foreign producers, and are likely to suffer. The agricultural export sector is one such example, and this mechanism is partly responsible for the observed tendency of traditional activities such as fishing and agriculture to stagnate in a petroleum economy. There may result an increase in the propensity to import that, given the inertia in consumer behaviour and in commercial trading patterns, will become difficult to reverse when conditions change. The non-tradeable sector may benefit from the increased demand that cannot be satisfied from imports, but as the supplies of non-tradeables may expand less rapidly than demand this may be a source of inflationary pressure. In monetary terms, inflation is accelerated if the additional petroleum revenues, which accrue in foreign currency, are monetised, i.e. if there are created equivalent monetary claims that are denominated in local currency. The sharp differences in rates of inflation experienced by major oil exporters in the 1970s can be largely traced to the way their monetary authorities handled this issue.

A second major influence is the tendency for the public sector to grow at the expense of the private sector, or, more correctly, for the administrative sector with its array of supporting services to grow at the expense of the goods-producing sector. This stems from the fact that most of the petroleum rent is appropriated by the government, which is likely to distribute it according to a political and administrative logic, with a large discretionary element. In principle, these funds could be handled in various ways: they could be used to reduce taxes, for example, or held in the form of foreign assets until suitable investment projects adequate to absorb them have been devised. In practice, a number of institutional and political pressures are likely to exert certain biases on expenditure decisions. In respect of their timing, there are always powerful reasons for spending now rather than later. As regards the type of spending, it is common for the size and/or levels of remuneration of the public service to be increased. The military services and the state petroleum enterprises are natural beneficiaries, but the precise pattern depends on local factors, including the particular distribution of political influence prevailing in the country. Whatever the economic or other reasons for supporting particular programmes, units or individuals, this nourishment of institutional interests may lead to rigidities which, when harder times come, limit the government's available options for restraining expenditure.

The tendency for civil servants to be urbanites means that enlarging the government apparatus often involves favouring the urban areas at the expense of the rural areas. The resulting changes in expenditure patterns may reinforce the tendency (based on exchange rate appreciation) for the import propensity to increase. The perception that the cities are where money and growth are to be found may act as a magnet drawing (particularly young) labour from the farms and rural areas. Particular temptations exist to hold down food prices in response to pressures from urban consumers. In combination, these factors may have an adverse effect on that part of the agricultural economy which produces for local consumption, as well as adding to the pressure, already mentioned, on the agricultural export sector. Against that, the increase in effective demand for food and agricultural goods should benefit local producers. There is evidence, for example from Ecuador (Commander and Peek, 1983), that it does so.

There may also be pressure to hold down fuel prices, often on the basis that the country's petroleum patrimony should be available cheaply to its citizens. (As we have seen, in Ecuador this principle was enshrined in legislation.) The resulting increase in local petroleum consumption (as in Ecuador, Egypt, Pakistan, Thailand, Tunisia) can render short-lived, or of limited impact, the hard-won improvement in foreign trade accounts achieved through petroleum production. In the case of both food and fuel, it is easier to initiate price subsidies in the buoyancy of expanding production and revenues, than it is to curb or eliminate them when it becomes necessary to do so.

Whether, when and how the (mainly positive) static effects will be superseded by the (largely negative) dynamic ones will depend on particular circumstances. The larger the scale of the petroleum sector relative to the rest of the economy and the more rapid the growth in its output and revenues, the more powerful the dynamic factors will be. The experience of several exporters (including Ecuador, Malaysia and Mexico) in the 1970s was exceptional, because sharp rises in their volume output coincided with dramatic price increases. It is no surprise that questions have sometimes arisen about whether the resulting revenues were always managed as wisely as possible. In the weaker petroleum markets of the 1980s, these pressures are much reduced. Most of the NODCs are not substantially affected by these dynamic influences because of the petroleum sector's modest share of their economies. However, since most of them aspire to expand that share, it is as well to be aware of some of the pressures that will be generated, usually only when they move beyond self-sufficiency to become substantial net petroleum exporters.

In addition to the other problems associated with a petroleum sector that is large relative to the rest of the economy, there is a difficulty of vulnerability stemming from the dependence involved. Given the inability of any individual producer to control the international oil market, sharp changes in demand and/or price can lead to dramatic changes in government revenue, export earnings, and economic activity levels (El Mallakh et al., 1984).

It has been convincingly argued that the conventional ways of measuring the petroleum sector's share in oil-producing economies are distorted, because they count as income (earnings derived from depleting petroleum reserves) that which is properly regarded as liquidation of capital (Stauffer, 1984). By adjusting the usual GDP figures for the static value of the exhaustible resource (by deducting a computed value of rent from current petroleum "income") and by making a "dynamic adjustment" for the multiplier effect that is generated when the oil rent is spent in the non-oil part of the economy, Stauffer arrives at an estimated "oil-independent" or "rent-independent" GDP for six producing countries of different types. This adjusted figure differs from the conventional GDP in proportion to the extent to which the economy is oil-dominated. Although none of the six is a NODC, his results are sufficiently germane to this study to present them here in summary, graphic form (see figure 5).

The implication of this analysis is that excessive dependence on petroleum – with all that that implies in terms of vulnerability to external price and demand changes, political pressure and an unbalanced economy – may appear much earlier (that is, at lower levels of production) than would be supposed from looking only at the conventional national income aggregates. This consideration should heighten the concern that has led many analysts to consider what kind of *depletion policy* might be advisable for various kinds of petroleum producers.

The debate about the theory and practice of voluntary restraints on petroleum production levels, in pursuit of various policy goals, has been long and complex, and it is clearly pertinent to any examination of how to maximise the benefits and minimise the problems of petroleum production; whether it yields any clear and helpful policy insights is less certain. Production restraints for other than technical or commercial reasons have been advocated or practised in the past for a number of quite distinct purposes:

(a) to maximise the discounted present value of all future production of the exhaustible resource. The most elegant and intricate theoretical discussions found in the literature are concerned mainly with this issue. However, the practical value of these analyses is severely limited because of inescapable uncertainty about values to be assigned to key parameters such as future prices and discount rates and technological developments affecting inter-fuel substitution (see, for example, Dasgupta and Heal, 1974).

(b) to shore up oil prices in the near to medium term. The cases of Saudi Arabian and OPEC production ceilings, and supportive measures taken by countries

Figure 5. Inter-country comparison of GDP impacts

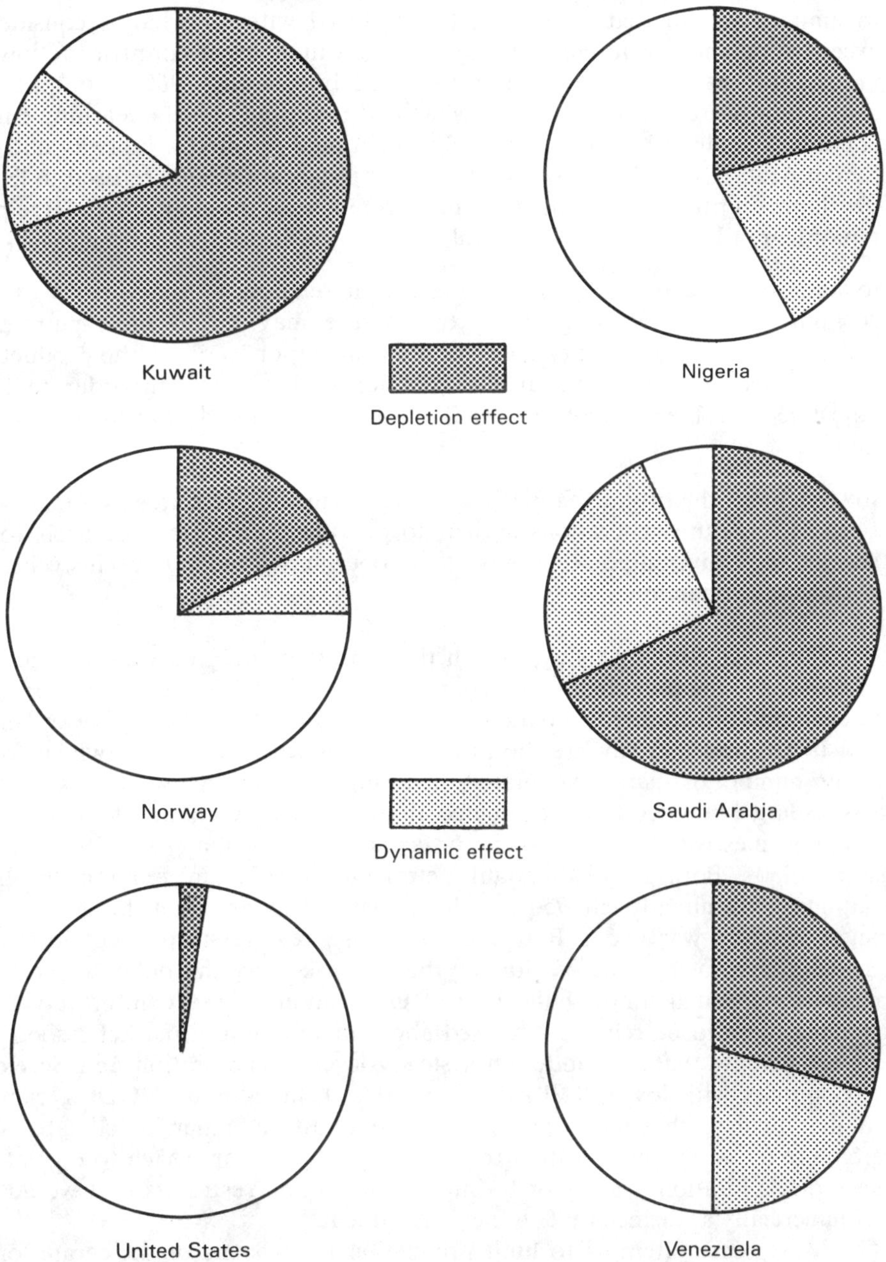

Source: Stauffer, 1984.

such as Egypt, Malaysia and Mexico, are a case in point; so is the activity in the United States earlier this century of the Texas Railroad Commission.

(c) to limit output to that which can be produced without what is considered excessive reliance on foreign participants. In a number of countries following nationalisation – e.g. Mexico after 1938 and Iraq in the 1960s – output and exploration were cut back to levels which, at that point on the learning curve, could be sustained by recourse mainly to national resources. Norway and the Syrian Arab Republic at various times in the 1960s and 1970s refused to let oil development proceed according to purely commercial momentum, in large part out of fear of losing national political control over the sector.

(d) to limit the size of the petroleum sector in relation to other parts of the economy. More specific goals of this nature may be to limit petroleum revenues to the absorptive capacity of the country; or to phase the production of a limited petroleum resource over a longer life cycle, in order to ease adjustment problems associated with the build-up and decline of petroleum output.

(e) to stretch out the producing lifetime of the exhaustible resource as a matter of fairness to future generations and/or to provide time to lay the basis for a sustainable, diversified economy that does not depend exclusively on petroleum production.

For most NODCs today, and in the context of this report, only the last two of these have the potential to be major issues. Those who wish to consider such policies will find only limited guidance from the experience of others. Kuwait (since the mid-1970s) and Norway are the principal examples of countries which for a significant number of years have limited their output to well below their capability for reasons largely related to the estimated ability of their economies to absorb the resulting revenues without major social or economic distortion (Atta, 1980; Ager-Hanssen, 1980). Both have abundant petroleum reserves in relation to their populations; both already enjoyed a high standard of living when the decision to restrict production was taken. Both have very high reserves-to-production (R/P) ratios – although in the case of Norway this is masked by the policy of curbing output partly through limiting the rate of exploration. These comfortably high ratios allow them to be relatively relaxed about maintaining a market-responsive investment climate that can underpin a steady level of exploration and development. Countries with lower R/P ratios, and this includes most NODCs, cannot afford that luxury; they must pay more heed to the commercial climate, and therefore are often not in a position to take a "fine-tuned" approach to regulating the level of exploration activity or to impose production restraints that would be seen commercially as disincentives to new investment.

Most other attempts to limit production for reasons of deliberate long-term strategy have been short-lived or half-hearted. As noted in the preceding chapter, Malaysia imposed such a limit for a brief time in 1980-81, although it was quickly set aside when revenue needs became pressing. Iraq limited its output to well below capacity during the period 1973-78, when rising prices allowed it to cover the needs of its development plan from a reduced volume of production. In Mexico (as in Ecuador) in the 1970s and 1980s there has been a lively debate about the macro-economic consequences of various levels of oil production and export. The 1980 decision to limit exports to 1.5 million b/d came under pressure after 1982 because of the need for additional revenues to service the country's external debt, exacerbated by weakening oil prices; but the ceiling was observed through 1985 and even, from late 1984, reduced to 1.4 million b/d as a gesture of support for OPEC. Continuing investment in additional production capacity, however, as well as growing revenue needs, suggested that this limit might be set aside in the later part of the decade.

The theoretical appeal of a long-range depletion policy based upon the desired relation between the petroleum sector and the rest of the economy remains considerable. However, the practical difficulty of devising and applying such a policy over a significant period of time – marked by major uncertainties regarding geology, demand, price and commercial conditions – should not be underestimated. It would appear from the record that the countries with the best chance of doing so are those with petroleum resources that are abundant relative to the size of their population and economy. By and large, the countries dealt with in this report do not have that characteristic.

Chapter 5

Petroleum and economic development

Most countries which possess petroleum resources seek to exploit them in ways that yield not only financial benefits but that also further the country's economic development in more direct ways. The general wish to keep at home as many as possible of the benefits to be derived from their resource often goes hand in hand with other, more specific motives:

(1) For countries that are relatively richly endowed with petroleum and that have few other natural resources, it is a question of "sowing the oil", to use the apt phrase coined in Venezuela – that is, of using the exploitable lifetime of their non-renewable resource to lay the basis for an economy that can be sustained when the petroleum is depleted.

(2) Or the dominant motive may be a wish to capture as much as possible of the value-added associated with petroleum activity; the concern here is a special case of that which is found in most countries that rely on the production of primary commodities.

(3) There is often a strategic consideration: given the high stakes associated with petroleum – in terms both of finance and of the economy's dependence on adequate energy supplies – governments of producing countries will usually wish to have and to be seen to have a voice in the key decisions regarding how their resource is developed. To be effective, this usually requires active involvement in the activity, which in turn requires a certain degree of competence with respect to the technology involved.

(4) Finally, for reasons of social policy it may be desired to avoid the kind of dualistic, enclave social and economic pattern that has often characterised the extractive industries in developing countries, and that means taking steps to integrate the sector into the local economy and society.

The basic characteristics of the industry make it more difficult to achieve these goals than to formulate them. Oil and gas development is extremely capital-intensive, and generates directly relatively few jobs. As most developing countries face capital scarcities and labour surpluses, this mismatch of resources must be dealt with. Much of the petroleum industry's activity is also demanding in terms of the level of technology required. This poses a challenge for those who wish to integrate it into their economy by, for example, supplying its required inputs or processing its output.

The specific goals in this sector that are appropriate for a particular developing country depend on a number of factors, principally *(a)* the size of its economy and industrial base and *(b)* the scale of its petroleum resource. The former bears on its ability to equip itself over time with the required industrial capabilities that are competitive in terms of cost and quality. The latter – the size of the petroleum resource – determines the life span over which that investment in human and industrial resource development is to be amortised. A commitment to national petroleum autonomy (in industrial terms) that would make sense for a resource base adequate to support production at a level of a million b/doe over, say, 50 years would be much more questionable if the resources are expected to be exhausted in 20 years. To keep the capabilities occupied by relying on foreign markets may not be impossible, but it would be foolhardy to count on this possibility; much better to justify them in terms of the scope of the national requirement.

Another factor that is relevant is the expected future behaviour of hydrocarbon prices. Production of some of the lowest cost oil and gas sources in the world was severely limited in the early to mid-1980s because of institutional and political factors, notably OPEC pricing and production policies and the war between the Islamic Republic of Iran and Iraq. Higher cost production, including some from NODC areas, has taken its place. If OPEC-related prices continue to drift downwards, in a world which for several years will probably have excess production capacity in low-cost areas, then pressure will increase on the higher cost producers. Even though governments in those countries will tend to protect local producers because of the longer term benefits that their survival is thought to bring to the local economies, the cost of this protection – and hence the pressure to reduce it – will grow if the prices of internationally traded petroleum continue to decline, so that they increasingly undercut more expensive production from national sources. In these circumstances not only is the (more expensive) local petroleum production itself economically vulnerable. *The forward- and backward-linked industrial activities, and the investment in human resource development, that are associated with that production may also be at risk.* We have seen in Chapter 3 a number of examples of petroleum-related industrial projects whose economics have become more questionable in a world of weaker prices and excess capacity. These were mainly downstream projects, especially LNG and petrochemicals. The same could occur, although the negative financial consequences may not be as great, for investments in developing suppliers of goods and services to the industry and even for investments in human capital.

This is a consideration that is worth bearing in mind by new or expanding producers contemplating the kind of industrial support effort that they should be making. The point is stressed not to deny but to put in perspective the positive returns that are normally to be expected from the development of indigenous petroleum support capabilities. Those positive returns are explicit or implicit in most of the discussion in the remainder of this chapter.

That discussion focuses on the question which is at the core of this project: what is the relation between petroleum production and economic and social development taken in a broad sense? As indicated in the preceding chapter, the emphasis here is not on financial or energy supply contributions of petroleum production. Important as these are, they are relatively well understood and documented. The main concern is rather with the petroleum industry's impact on those basic determinants and/or indicators of development: human capital formation, industrialisation, and the spatial distribution of infrastructure.

Employment and manpower considerations

The human resource issues of greatest interest in relation to this study are the following:

- What are the employment and labour market effects of petroleum activities, especially in the upstream? and
- How can a developing country with relatively little upstream petroleum experience go about acquiring the skills needed to manage its own petroleum sector? What is involved in an indigenisation programme?

These will be explored in turn.

Employment generation and labour market effects

Direct job creation is not a primary reason for seeking petroleum. By far the greatest contribution that petroleum production makes to employment levels is by easing or removing the balance-of-payments or energy supply constraint on economic growth, thereby making it possible to generate more jobs throughout the economy. The oil and gas industry in general requires relatively low numbers of staff, and the upstream parts of the industry employ even fewer people on a worldwide basis than the marketing – distribution – administration segment (see table 4).

It should be noted, however, that the absolute numbers employed are not negligible, especially if indirect effects are taken into account, and that in qualitative terms the petroleum sector workforce, especially in countries where there are upstream operations, embodies skills and expertise of a high order. Although no satisfactory body of comparable labour statistics for the petroleum industry has been compiled, the following data and observations will help provide a perspective.

The relative importance of petroleum employment is greatest in countries with substantial petroleum sectors and small populations, yet even there it represents only a small share of the labour force. In Trinidad and Tobago about 4 per cent of the labour force is occupied in the petroleum sector. (This appears to include at least part of the petrochemicals sector. Turner, 1982.) In the major exporters of the Middle East the figure is lower. Petroleum employment in the member countries of the Organisation of Arab Petroleum Exporting Countries (OAPEC) amounts in the aggregate only to about 200,000, or around 1 per cent of the working population of those countries. (This excludes Egypt. Norconsult, 1984). Of the countries examined in Chapter 3, China has the largest petroleum employment, officially put at 633,000, including around 200,000 women. (The latter figure may be accounted for by the fact that many support services – catering, housing, education, health and so on – that in other countries are supplied by contractors are internally provided in the Chinese industry.) This figure may not include all exploration workers and almost certainly does not include all refinery

Table 4. Estimated past and future employment in the world petroleum industry: Statistical data and estimates for 1960, 1980 and 2000

Country groupings	Estimated numbers employed in the petroleum industry			
	Exploration and production ('000s)	Refining and blending ('000s)	Marketing/ distribution/ administration/ support ('000s)	Total ('000s)
1960				
OECD	380	280	760	1 420
OPEC	130	35	24	190
LDCs	140	80	115	335
CPEs	180	65	140	385
World	830	460	1 040	2 330
1980				
OECD	640	300	725	1 665
OPEC	150	45	75	270
LDCs	140	115	290	545
CPEs	270	110	450	830
World	1 200	570	1 540	3 310
2000				
OECD	640	170	525	1 335
OPEC	170	60	155	385
LDCs	190	95	275	560
CPEs	300	115	435	850
World	1 300	440	1 390	3 130

Source: Adapted from MacKay, 1983.

personnel, but even if one doubles it, it represents only a fraction of 1 per cent of the country's total workforce (Hills, 1985).

These employment levels may be significant not only because of their magnitude but also because of their distribution. In particular, the concentration of employment in large state petroleum companies is a phenomenon that may have local significance in terms of its political and economic influence. The practices of these enterprises in matters of industrial relations, wage rates and working conditions may be emulated by other local companies, with effects that can be both positive and negative. Of the countries under review, the following figures will convey an idea of the significance in terms of employment of some of these companies.

Just as employment may be distributed in a lumpy fashion as between firms, it may also be unevenly distributed over time. Some phases of the integrated petroleum chain are more labour-intensive than others. In particular, the construction activity associated with field *development* requires a relatively heavy labour input but for a comparatively brief period, say two years for a field with a producing lifetime of 25 years (see figure 6). Hence a country whose level of

Figure 6. The phasing of petroleum-related employment (offshore)

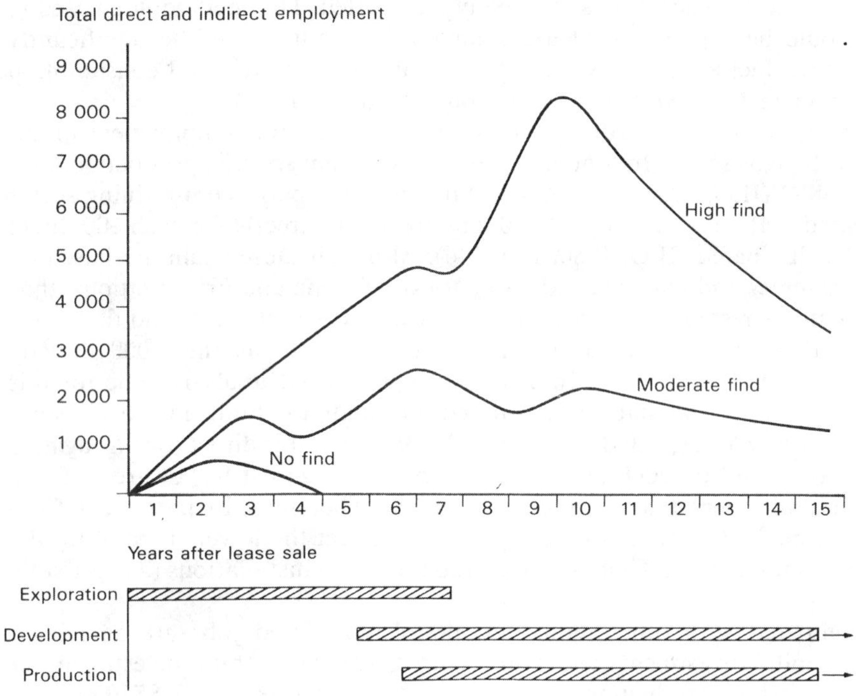

Source: Gilbert, 1983.

upstream petroleum employment is relatively stable over the years must have shifted a significant part of its labour force from one development site to another over time, and this implies a lot of geographical mobility and/or on-site recruitment for the workforce.

The indirect employment generated by petroleum activity is even more difficult to estimate than the direct, but some studies have been made which give an idea of the order of magnitude. One of the more carefully done examines six large, petroleum-related investment projects in the Netherlands: two gas processing units, two refining units, one offshore installation and one petrochemical plant. The indirect effects are broken down into those associated with the investment itself (intermediate inputs needed to produce it), those associated with its operation (inputs needed for running the installation during production phase), and a "consumption effect" stemming from the additional disposable income due to the investment and production effects. It is found that at the investment stage, each million guilders invested generated slightly more employment indirectly than directly (4.1 work-years indirect investment effect plus 5.4 work-years consumption effect, as against 7.4 work-years directly). At the production/operation stage, it was found that the different investment projects created between 1.9 and 3.6 jobs outside the company, through a combination of consumption and indirect production effects, for each regular company job created. The methodology used in that study could be applied elsewhere, although the results would be significantly affected by such factors as real wage rates, labour productivity, and consumption and import propensities (Netherlands Economic Institute, 1984).

Other, less precise studies have estimated the indirect employment multiplier to be as high as seven (Indonesia) or even, for certain specific production jobs, 20 times (USSR) (ILO, 1986, III, pp. 4-7). Studies of the petroleum-refining sector have estimated that between 190 and 360 jobs are lost indirectly for each 100 direct refinery jobs eliminated (ILO, 1986, I, p. 146). Although these again are based on European examples and may need adapting for developing country situations, they are in line with the results of another study on Aruba, where it was found that when the refinery there was closed, 3,750 indirect jobs as well as the 1,000 directly involved were lost (Haan, 1985). In Thailand input-output analysis done for this project showed that the natural gas-based installations built in the Eastern Seaboard Project were expected to generate by 1990 3,000 indirect jobs as against 4,600 direct ones and by 2000 about 7,000 indirect as against 9,500 direct. These imply lower multipliers than in the cases above. That may in part stem from differences in methodology; however, it may also reflect the lower share of locally manufactured inputs in the Thai as against the European installations (Tingsabadh, 1985).

Compared with other industries, petroleum-related jobs are associated with higher capital investments and skill requirements. The capital investment per worker on an offshore production platform may easily approximate US$10 million. In the case of a modern refinery with secondary and cracking capacity, the figure may be around 1 million (Shell Internationale Petroleum Maatschappij, 1980). As regards skill levels, it was estimated in 1980 that the share of professional-level personnel (technical, administrative and managerial) in a cross-section of the international oil industry was above 20 per cent and rising (see figure 7) (see also Zymelman, 1980; World Bank, 1980).

Because petroleum industry cost structures are dominated by capital expenditures, with wage costs of secondary importance, and because skill requirements put a premium on having well-qualified employees, compensation levels in the industry have tended to be above those in most other sectors and working conditions comparably superior (ILO, 1977). While this has resulted in a lower incidence of industrial relations problems than in some other sectors, it has sometimes created problems through a spillover process. One form this can take is a process of wage competition, whereby wages in non-petroleum sectors where productivity is lower get bid up to levels that make those sectors non-competitive internationally. This has been a problem, for example, in Trinidad and Tobago (World Bank, 1983b). The wage cost pressure can also be transmitted via certain professional specialities that are required in other sectors as well as petroleum. This occurred, for example, in Indonesia, where the oil boom put pressure on the market for such qualified personnel as engineers, accountants and mid-level technicians (ILO, 1986, III, p. 26).

Figure 7. Employees in the world petroleum industry – estimated changes in occupational categories, 1960-2000.

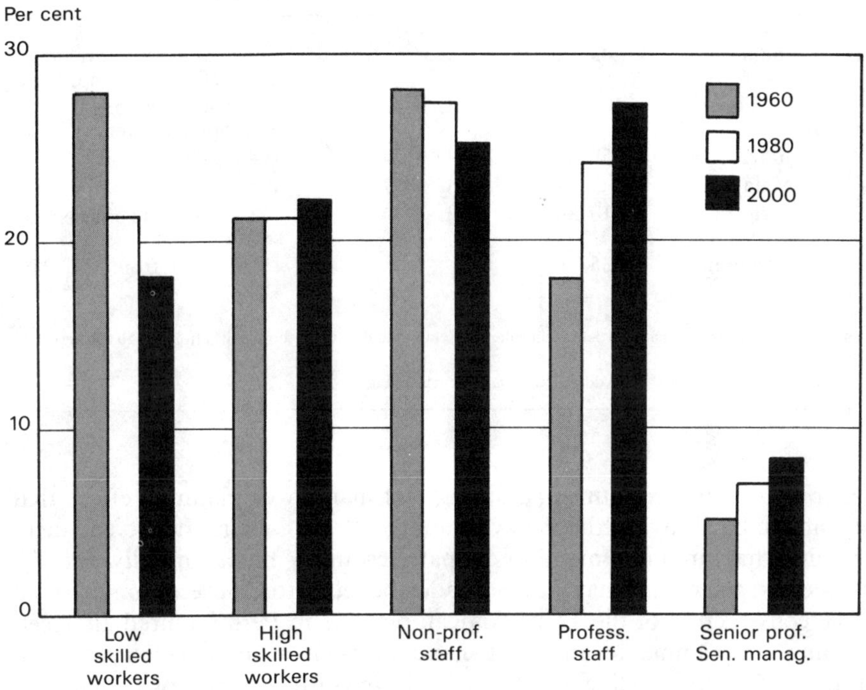

Source: MacKay, 1983.

The difficulty may take another form in countries where a significant part of the petroleum industry is in the public sector (see table 5). There, pressure may be felt to keep salaries in the state petroleum enterprise or enterprises at levels applicable to other civil service or public sector employees. Given the generally higher compensation available elsewhere in the petroleum industry, this may lead to a loss of key personnel to private or foreign firms. This occurred, for example, in Peru in the 1970s, when salaries were held down in Petroperú in order to keep them comparable to scales in other parts of the public sector. Since the same restraint was not applied by private petroleum companies in Peru or by petroleum firms outside the country, skilled employees with long experience were bid away in significant numbers, with damaging effects that have taken the company many years to correct.

One notable feature of petroleum-related jobs is that relatively few of them are specific to that sector. Most require a technical ability – pipefitting, welding, electricians, accountants, computer science, etc. – which is transferable between sectors. Table 6 shows the approximate distribution of skills in a typical upstream company. This feature of petroleum-related employment gives much-

Table 5. Employment in state petroleum enterprises

Company	Country	Employees	Sectors [1]	Comments
ONGC/OIL	India	43 000	U	Contract workers excluded; about 80 per cent in ONGC
Pemex	Mexico	103 000	U + D	1979. Includes 45 000 temporary workers
Petrobrás	Brazil	50 000	U + D	1983
Petronas	Malaysia	5 700	U + D	1985
Petroperú	Peru	10 000	U + D	1985. 25 000 in Peru's oil/gas sector
YPF	Argentina	35 000		1979. Down from 53 000 in 1976

[For comparison, Exxon, the largest international petroleum enterprise, employed directly or through its affiliates 150,000 persons world-wide in early 1985.]

[1] U = upstream (exploration, production) D = downstream (refining, marketing).

needed flexibility to the recruitment, training and manpower planning effort that otherwise tends to be always at the mercy of unpredictable factors. It also enhances the probability that an investment in human resources made initially for the petroleum sector specifically may have broader benefits for the economy. As an example, the construction of the Trans-Andean pipeline in Peru required an accelerated training programme to supply a large number of Peruvian welders to construct that 856 km pipeline under difficult operating conditions and time constraints. Not all of them could be retained in the petroleum sector following this one-time project, but the skill was transferable to other sectors.

Training and indigenisation

The process of developing nationally based petroleum expertise in NODCs has been under way for many years and is now well advanced in most countries where there is significant oil and/or gas production. There are many dimensions to this effort, including on-the-job training, the development of local training and educational institutions, scholarships to specialised institutions or programmes in other countries, joint ventures with foreign companies in which they undertake to train local staff in their own operations, international cross-postings for more senior personnel to help keep them abreast of state-of-the-art technology, etc. (ILO, 1981). It is appropriate to initiate and shape this programme according to the particular country's educational attainments at the time that it embarks on a major upstream petroleum commitment. Workers' representatives often have a useful role to play in helping identify training needs and shaping appropriate programmes to meet them. (Conclusion No. 71 of the ILO Petroleum Committee – at its Ninth Session in 1980 – called on governments to develop training policies for the petroleum industry "in consultation with representative employers' and workers' organisations".) Provisions for transfer of technology in general and training and employment of local staff in particular are nowadays common in contractual arrangements governing exploration and production in NODCs (United Nations Centre on Transnational Corporations, 1983). Even in the 1980s, when many countries were making their terms more attractive in an effort to lure investors who had become more hesitant, these provisions continued generally to be insisted upon. However, China, whose terms on these issues were very demanding in the first years after the opening of offshore activity to foreign companies, later eased its requirements in view of the friction experienced with foreign operators over these matters as well as the disappointing exploration results and the weakening oil market (*Petroleum Intelligence Weekly*, 16 December 1985). Tables 7 and 8 summarise the operation of the indigenisation process in two unidentified developing countries.

This process can and no doubt will be taken further, as the technological frontier dividing those tasks which are performed by local staff or enterprises and those which are left to expatriate individuals or firms continues to be pushed back in existing producing countries, and as new or expanding producers embark on the process of indigenisation that their predecessors have already followed. There will continue to be questions about how far a particular country can or should go in equipping itself with an autonomous petroleum capability, but on its right to move in that direction there is no longer any quarrel. Exxon now points out that over 98 per cent of its employees world-wide are nationals of the countries in which they work (Exxon Corporation, 1985). A similar situation prevails in Elf-Aquitaine (Rapport social, 1983). And these are not isolated examples in the industry. Arguably there is no advantage to be gained by anyone if such firms aimed at a 100 per cent figure, for there is value to the various parties involved in the spread of technology and experience that comes from international cross-postings. While companies continue to seek international experience for a core of their staff, a recognition of political realities and the cost savings of using local staff rather than expatriates will reduce to a minimum the number of situations where companies and host countries differ over staffing particular posts.

In the mid-1980s, when employment in the industry was contracting in the OECD countries and expanding in many NODCs, some developing countries could see advantages in recruiting for their own account foreign individuals who had petroleum industry experience but who had become free agents, as a cost-effective way to assist in training their own cadres and building their own institutions. This was not a new option, but by then it aroused fewer suspicions than formerly, because of the changed roles of companies and state operators in the industry and of the mutual education process that had taken place.

Many NODCs have gone quite far in developing indigenous petroleum expertise. India and Mexico staff virtually the whole of their industry with nationals and rely on foreign contractors for a very small share of total operations. In Egypt only a very small number of expatriates are involved. In Tunisia, a relatively small and recent producer, the largest producing field, accounting for over half the country's output, is operated entirely by Tunisian personnel; the project director for an extension of the second largest field, which is offshore, is Tunisian. Some countries have covered this distance mainly by their own efforts

Table 6. Typical skill distribution of an oil and gas exploration and producing company

Oil related [1]			Technical support [2]	
Scientific	Engineering	Prod./drilling petr. eng.	Technical support	Operations support
Exploration management	Engineering management	General management	Technical assistant (financial)	Operator
Interpretation geologist	Electrical/ instruments (and design)	Reservoir and production engineering	Technical assistant (engineering)	Multi-skill technician
Interpretation geophysicist	Process (and design)	Evaluation and general petrol engineering	Computer operations	Technical assistant (production)
Other geol-ogists and related	Civil and structural (marine and naval)	Petrophysics	Technical assistant (exploration)	Safety (and medical)
R & D (engineering)	QA/QC	Drilling/ well site	Draughting	Marine services
R & D (exploration)	Costs/contracts/ planning	Platform engineer (electrical)	Technical assistant (petroleum engineering)	Operations laboratory
	Mechanical (and design)	Maintenance (electrical)	Stratigraphic laboratory	
	Telecoms.	Maintenance (mechanical)		
	Services (applications) systems/services	Maintenance (other)		
	Projects (construction)	Terminal installation management		

[1] 24 per cent of staff. [2] 32 per cent of staff. [3] 18 per cent of staff.
Source: PA Management Consultants, 1985.

after nationalising the industry (Mexico and to a large extent India). Others have taken a more evolutionary and co-operative approach with the former concessionaires (Nigeria, Venezuela) (A. Parra, 1977). To convey more of a sense of the quality of this process, we will go into greater detail on three countries, one each in Latin America, Africa and Asia.

Brazil. In Brazil the training effort has been directed by Petrobrás. From the very early days (1950s) it was decided to emphasise the development of training institutions in the country and to minimise the sending of Brazilians abroad for training. As a result, a significantly smaller fraction of Petrobrás professionals has been trained in the United States than is the case for their Venezuelan counterparts in Petrovén (less than 1 per cent as against 4-5 per cent). One of the first steps was to institute a programme of instruction in geology at the University of Bahia, near

	Professional support[3]			Administration
ıpply and ades	Computer	Personnel comm.	Financial	
upply ontracts	Exploration data processing	Personnel	Financial manager	Administration
Iaterials ontrol	Seismic/non-seismic data processing	Partnership and planning	Financial accounting	Property and office services
uying/ urchasing rocurement	Computing management	International management	Cost accounting	Clerical
Varehousing nd transport	Project systems	Legal	Audit	Secretarial
killed radespeople/ rivers	Finance and administration systems	Taxation	Financial planning/ analysis	Information retrieval/ librarian
ecurity	Technical computing	Contracts and negotiations	Financial administration	WP/VDU computer terminal operators
hauffeurs nd drivers	Support services	Treasury	Project/ management accounting	Telephone/ telex operators
		Corporate planning	Economic/business analyst	Press and public relations
		Operational planning	Systems accounting	Printing/ publishing/ photography

the then-known petroleum deposits, with the help of an instructor on loan from Stanford University. The early crop of exploration and production specialists trained there subsequently rose to senior positions in the company. Many of the Brazilians who did go abroad for training went during the early years and returned to teach in Brazil.

The emphasis of the training effort changed over time as a function of conditions in the industry and in the educational system. The early stress was put on exploration and production, but following the disappointing results of the onshore search and a switch of the company's priority towards the refining sector this change was reflected in the training programmes. More recently, as interest shifted to upstream activity offshore, the emphasis has reverted to exploration and production. As Brazilian universities developed their offerings in petroleum-related

Table 7. Example of the training of nationals in a developing country by a multinational oil and gas company

Year	Company action/event
1962	*Prospecting licence granted*
1962	Nationals sent to company's country of origin for training in: petroleum geology drilling administration finance University scholarships awarded abroad
1963	First local training centre established comprising drilling rig and workshops for training: drilling personnel mechanics electricians welders fitters
1965	*First oil discovery*
1972	Expansion of local training facilities for the training of basic oilfield skills
1972-85	Expatriates gradually replaced following a Training and Development Plan Long-term employees received training in specialised technical areas of exploration, production, maintenance, administration and finance As the employees developed, courses in managerial skills were introduced
1985	*Today the majority of the senior positions are held by nationals*

Source: PA Management Consultants, 1985.

disciplines, a diminishing share of the training was carried out by Petrobrás itself. Similarly, as general educational levels rose, the company was able to discontinue some of its lower level instructional programmes; for example, those in literacy training.

The country's national industrial apprenticeship system, SENAI, requires firms to pay an amount equivalent to 1 per cent of their payroll expenses (1.2 per cent for large firms) to a training fund, but up to 80 per cent of that may be recovered for internal training expenses through agreements negotiated between the firm and SENAI; this has covered an important part of Petrobrás's training costs. Similarly, accords with several of the country's universities have given Petrobrás the flexibility of expanding or contracting the output of various specialities according to need without the burden of creating a permanent internal capability to train large numbers of experts in all specialities. An effort has been made to relate the production of various specialists to national manpower requirements so as to avoid problems of overproduction, as well as those of shortages, of highly skilled personnel. For this purpose manpower planning projections are carried out by Petrobrás through its Human Resources Management department.

Table 8. Assistance given by an international company in a developing country over a period of nine years, and resulting level of indigenisation

Action/result	1	2	3	4	5	6	7	8	9
Formal company training (local and overseas)	626	790	2 420	2 333	1 788	2 514	1 945	3 826	2 757
Work-days of formal training received both locally and overseas	29 200	30 253	52 737	36 209	30 416	45 423	38 594	33 657	49 436
Scholarships									
Universities	131	137	137	119	88	73	47	34	10
Technical colleges	60	49	44	30	16	16	12	12	–
Marine scheme	8	8	8	12	12	12	7	7	7
Manpower summary									
Indigenous staff	1 340	1 920	2 620	2 900	2 960	2 930	3 040	3 150	3 280
Expatriate staff	320	500	480	400	320	230	220	190	180
Total	1 660	2 420	3 100	3 300	3 280	3 160	3 260	3 340	3 460

Source: ILO, 1986, III, p. 56.

From 1953-83 some 151,690 work-courses at mid-level and 60,369 at high level took place; on average, each Petrobrás employee during that period participated in four personnel development projects. As of 1984, 3.2 per cent of the company's work-hours were spent in training activities. The overall training effort from 1973-83 cost 10,958 million cruzeiros. The greatest share of expense went to mid-level training, under accords with SENAI. The total number of foreigners employed by Petrobrás from 1954-84 was 381, most of whom were hired either during the early years (1955-64) or since the mid-1970s, as the emphasis shifted to offshore activity. (This section relies heavily on notes of Professor Laura Randall for a forthcoming book on the petroleum sectors in Brazil, Mexico and Venezuela.)

Côte d'Ivoire. A notable feature of the Côte d'Ivoire's petroleum experience is the progress made – in a relatively short time and on the basis of modest reserves – towards staffing the sector with trained nationals of the Côte d'Ivoire. At the time when a State Secretariat (later to become the Ministry) of Mines was created in 1971, there were only three geologists from the Côte d'Ivoire. Almost immediately, a programme was initiated of carefully selecting, recruiting and giving university training in the earth sciences to young nationals of the Côte d'Ivoire. Under this policy 255 trainees had by mid-1985 undergone formal studies in these fields, 113 at institutions within the country and the remainder at a variety of universities and advanced training institutes in North America and Western and Eastern Europe. Around 100 of these were at that time working in the country's petroleum industry or corresponding sections of the civil service.

In addition, substantial training activities were conducted by or through the operating oil companies. Phillips, the company with the most extensive upstream operations, spent between 1981 and 1985 approximately US$4 million on training activities – vocational, on-the-job, safety, scholarships, outside training and various assignments for personnel of Petroci (the state oil company of the Côte d'Ivoire) to Phillips' affiliates in the United States and the United Kingdom. It arranged overseas training for 11 Petroci staff during that period – eight engineers and two geologists in the United States, one geologist in the United Kingdom. It selected and trained 48 supervisors *(agents de maîtrise)* at its own expense, then selected and employed 19 of them. As of mid-1985, 117 of the company's 157 employees were nationals of the Côte d'Ivoire. Of the 40 expatriates, five were scheduled to be replaced by nationals of the Côte d'Ivoire. Professional posts held by nationals of the Côte d'Ivoire included the company treasurer, chief computer specialist, two engineers, one geologist and three accountants. Esso, which operates the smaller of the two producing fields, then employed about 85 persons in its upstream company, some 80 per cent of whom – including the Belier platform manager – were from the Côte d'Ivoire. The staff of Petroci numbered 246 nationals of the Côte d'Ivoire and five expatriates. Most of its 50 professionals were trained in the United States, smaller numbers in other countries such as Canada, France and Romania. By 1985 no new university students were being sent abroad for petroleum training because with the reduced level of activity it was already difficult to find places for those returning from earlier assignments.

The "Ivorisation" achieved in the upstream was preceded by a similar – if less difficult, because technically less demanding – process in the downstream.

Stimulated both by general factors related to cost and to political realities and, after 1978, by a formal government policy asking firms to prepare, on a non-binding basis, an "Ivorisation plan", and to report annually on its implementation, all downstream petroleum companies made considerable efforts in the late 1970s and early 1980s to replace expatriates by local staff. The situation in mid-1985 was as follows:

	Total employees	Expatriates
Mobil	200	5
Shell	180	4
Texaco	72	3
Esso		
downstream	not available	2
upstream	85	17 (approximately)
BP	140	3 – all leaving soon
AGIP	203	10 – only 1 of 8 departments headed by expatriate

A decade earlier, the expatriate numbers in these companies were typically ten times or so higher than these. A corresponding training effort was required to achieve these results. It was facilitated – as in the upstream – by a tax credit whereby up to half of the employment tax which companies must pay (1.5 per cent of their wage/salary expenses) is credited to them against training expenses incurred within the country for their employees. Much of the training related to refinery operations and safety questions related thereto was carried out at a very modern facility built by the refining company SIR for that purpose at Grand Bassam. However, while the facility itself was first rate, the requirement for such an investment was less clear; it has been underutilised and its fixed charges represented a continuing financial burden for SIR.

Malaysia. Indigenisation has been a primary objective here as in Brazil, with the additional goal of advancing *Bumiputras* within the Malaysian contingent. But the starting-point and the approach have been different, since the private sector plays a much larger role in Malaysia's petroleum sector. A larger part of the training effort is made by the private contractors and a larger part takes place abroad than is the case in Brazil.

Malaysia's production-sharing contracts contain clauses relating to the companies' obligations regarding Malaysianisation, *Bumiputra* participation and the transfer of technology. The execution of these commitments is systematically monitored by Petronas. Manpower forecasts are submitted by the oil companies as part of the annual work programmes, and their budgets for training have to be approved. Further controls are exerted through the procedure of issuing work permits for expatriates. The overriding objective is to enable Malaysians to attain a greater degree of self-reliance and increased participation in the petroleum industry. The petroleum companies are generally seen to have co-operated genuinely in this aim. Expatriate numbers have declined progressively and have now stabilised at a low level which may be the minimum compatible with an effective transfer of technology.

Joint ventures with the major oil companies have provided one of the principal means of getting the (on-the-job) training needed to achieve the objectives of indigenisation. The other avenue has been an extensive programme of formal training schemes. Training centres are operated by Petronas and Esso in Trengganu and by Shell in Miri (Sarawak) and Labuan (Sabah). Petronas staff are also seconded to the contractors' organisations, both in Malaysia and overseas, on assignments aimed at gaining work experience; in 1983/84 a total of 66 Petronas staff were so assigned. And extensive use is made of overseas scholarships. Budgets are relatively generous and numbers of trainees are impressive. Esso and Shell provided training programmes in 1984 totalling about 16,000 and 50,000 work-days, respectively. Petronas spent over M$20 million (US$8 million) on training from 1981 to 1984, benefiting 13,513 staff, or an annual average of nearly 4,000 (compared with a total employment in 1985 of 5,700). This does not include M$60.6 million spent over the same period on scholarships, which benefited 1,721 Malaysians. Most of these are at degree level, and most are in scientific and technical disciplines. These include scholars going overseas (predominantly to English-speaking developed countries) and to Malaysian and Singaporean universities. Private companies also award scholarships. Esso provides 32 awards per year, 187 Malaysians having benefited since the programme began. In the period 1981-85 Shell granted 85 overseas scholarships (76 to *Bumiputras*), 260 scholarships to Malaysian tertiary institutes (221 to *Bumiputras*), and over 500 bursaries to Sabahans and Sarawakans (practically all to *Bumiputras*).

The effect of this effort on Malaysianisation has already been referred to in Chapter 3. In the case of the two largest foreign-owned petroleum organisations, the results have been as follows. In Shell's upstream operations, total employment grew between 1973-84 from 693 to 2,871. During this time, expatriate employees grew in absolute numbers from 64 to 230, but as a share of senior staff their contingent declined from 72 per cent to 32 per cent, and was planned to drop to 24 per cent in 1986. *Bumiputras* have increased their share of the Malaysian management level staff from 39 per cent in 1980 to 42 per cent in 1984, and at the level of supervisory staff from 39 per cent to 50 per cent. For Esso, less detail is available, but in an enterprise (downstream as well as upstream) that expanded from 963 employees in 1975 to 2,488 in 1984, the share of expatriates rose from 11 per cent in 1975 to 17 per cent in 1979 and then fell back to 11 per cent in 1984. Petronas employed 34 expatriates, out of 5,742 employees, in early 1985 (section based on Bowie, 1985).

* * *

What matters more than the figures, of course, is the degree of competence and autonomy that are achieved as a result of these efforts of training and indigenisation. In the case of Petrobrás, by dint of great effort over three decades the company has now arrived near the forefront of world petroleum technology in its offshore development efforts in deep waters. In Malaysia's case, the state oil company is not the unique embodiment of national petroleum capacity in the same way. Two decades younger than its Brazilian counterpart, Petronas has none the less come rapidly up the learning curve. Through its exploration subsidiary Carigali it has been involved in joint venture operations with BP, Elf and Shell, all in offshore areas. But much of Malaysia's petroleum expertise is found in the local affiliates of international oil companies. The Côte d'Ivoire, where upstream skills are divided between the state company Petroci and the private operators Phillips and Esso, is an intermediate case. The two approaches to acquiring expertise have different advantages. There can be disagreement about which is more effective or preferable, but in these instances both have produced impressive results.

In many developing countries, OPEC as well as NODCs, an important test of their manpower development programmes is only beginning. Many of them indigenised on the basis of local staff who had acquired their training and experience in the private oil companies before nationalisation. As many of this group of countries are now a generation into the experience of running their own industries, the personnel who will hereafter be moving into senior posts will in those cases be the first cohort which has been trained entirely in an environment of local self-reliance.

Industrial spin-offs

As in the case of building petroleum expertise in the national labour force, the objective of capturing oil- and gas-related industrial benefits tends to be a secondary aim, one which follows and is incidental to the production activities. That is especially true with respect to the backward-linked activities that supply needed materials and services to the petroleum operators. In some countries, such as France and Japan, important industrial capabilities have been developed to support upstream petroleum activities, in the absence of significant indigenous petroleum production. But these are exceptions, as in general the existence of local petroleum production leads to the growth of the industrial activities that support it. In the case of forward-linked, processing industries, the picture is not quite so clear-cut. Refinery and petrochemical capacities have tended in the past to be located near their markets rather than the sources of their raw materials. However, in recent years the desire of producers to retain locally as much as possible of the value-added associated with the petroleum they produce, and their ability to price their feedstock competitively, have created a tendency for these activities also to migrate towards the sources of production.

Supply of materials and services

The scope that an individual country has for providing services, material and equipment to the petroleum industry depends, of course, on its size and industrial base. It also depends on the nature and location of the petroleum deposits, especially whether they are onshore or offshore, as the technology required in the latter case may be much more demanding. And capabilities can of course evolve over the producing lifetime of the deposits. At the low end of the scale, almost all countries will be able to provide construction, catering and trucking services. The moderately industrialised may be able also to provide onshore drilling services, steel pipe, helicopter services, some drilling muds. Only the more advanced and experienced – in our sample, Brazil, India and Mexico – may be in a position to supply some of the more sophisticated items, such as drilling ships, offshore platforms, "Christmas trees", high-pressure pipe and valves. Some sophisticated materials and services such as well logging are available from only a few sources in the world and are effectively beyond the reach of all NODCs at the present time.

Contracts nowadays commonly require companies to use locally produced materials and services, provided they are acceptable in terms of quality, delivery time and repair service. Price is also a relevant factor, but usually not the telling one; where acceptability on other criteria is established, companies will often be co-operative in procuring locally, provided the price penalty is not exorbitant. The key to trouble-free procurement decisions is regular communication between companies and government (or its agent, which may itself be a company, in the form of the state petroleum enterprise) in an atmosphere of mutual confidence, so that quarrels do not arise over what is "acceptable" or "exorbitant".

The possibilities and difficulties can be made clearer by reviewing the experience of a few countries.

Brazil. Reference has already been made to Petrobrás's role as stimulator of the country's capital goods industry, some 30 per cent of whose orders it accounts for. It was instrumental in establishing the Brazilian Capital Goods Industry Association in 1955. It provides firms in the sector with bankable long-term contracts on the basis of which they have been able to finance long-term product development. This has typically been done in joint ventures with foreign firms until the Brazilian partner has mastered the technology in question, at which time it assumes alone the responsibility for further contracts. This has been the route followed for example by a firm like Montreal Engineering. In 1983 Petrobrás established a "nationalisation division" with the purpose of supervising the development of new products and suppliers inside Brazil. In 1984 Petrobrás bought US$800 million worth of equipment, nearly 80 per cent of it from Brazilian sources. The learning curve can be graphically presented and closely follows the changing emphasis in the industry from onshore exploration to refining and finally back to exploration and production but in offshore areas (see figure 8).

Foreign contractors are expected to buy from Brazilian suppliers irrespective of price. According to cost comparisons provided by a major oil company operating in Brazil on risk contracts, Brazilian prices for commonly used drilling materials averaged 7.5 per cent higher than United States prices, as of early 1985, although prices of some of the 21 typical items compared differed from United States prices, up or down, by much wider margins. Not all of Petrobrás's import substitution programmes have been equally successful. The shipbuilding sector was promoted during the 1970s on the basis of prices over twice those available from external sources, and the capacity thus created had to face the slump of the 1980s with a high cost structure, heavy debts and low order books (section based on Gall, 1985).

Figure 8. Development of import substitution of capital goods for the petroleum industry

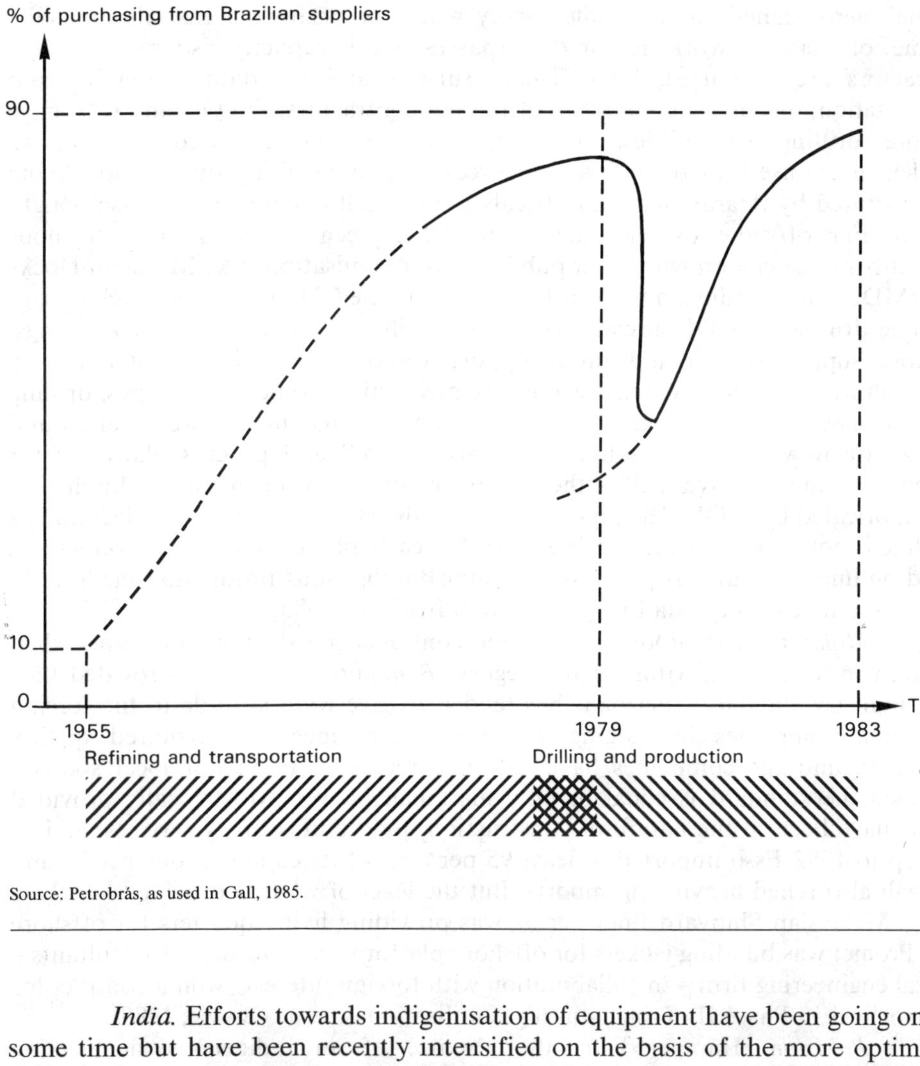

Source: Petrobrás, as used in Gall, 1985.

India. Efforts towards indigenisation of equipment have been going on for some time but have been recently intensified on the basis of the more optimistic outlook for reserves and production. A special government unit (Empowered Committee of Secretaries) and an "Indigenisation Cell" in ONGC and OIL have been established to that end. The latter disseminates information on the equipment requirements of upstream petroleum operators to the relevant parts of manufacturing industry. Joint ventures are also being encouraged between Indian firms, public or private sector, and foreign companies in the area of oilfield services, with the aim not only of supplying Indian needs but ultimately of exporting to other national oil companies and developing countries. (Reliance on joint ventures with foreign partners, as a way to stimulate the transfer of technology, is also an approach used in *China*.)

As of 1985, the division between locally and externally supplied equipment and services is as follows. Indigenous sources are available for gravitational, aeromagnetic and seismic survey work for onshore areas, although the volume of seismic work required surpasses local capacity, so that foreign contractors are also used. For offshore surveys and for data processing and interpretation, extensive use is made of foreign contractors. ONGC and OIL have onshore drilling capabilities, but again not in the required volume, so supplementary use is made of foreign sources. Onshore drilling rigs are now being manufactured by Bharat Heavy Electricals Limited, although some are also bought abroad. For offshore drilling, the strategy has been to build up indigenous capability in association with other public sector organisations like Mazagon Docks Ltd. (MDL) and Hindustan Shipyard Ltd. (HSL). The ONGC owns six jack-up rigs and one drill ship. MDL already has the capability to manufacture jack-up rigs, offshore supply vessels and multi-support vessels; and HSL is planning to manufacture drill ships. As regards onshore production, some casing pipes, drilling bits and natural gas compressors, and most low- to medium-pressure valves and fittings are now produced indigenously. So are well and process platforms for offshore production. Over half of the 48 offshore platforms installed by March 1983 were fabricated by MDL. The pipe required for the Hazira-Bijaypur-Jagdishpur gas pipeline is not now fabricated in India, so that early phases of construction will be based on imports, but it is planned to establish indigenous production facilities in 1987 (section based on Tata Energy Research Institute, 1985).

Malaysia. The production-sharing contracts specify that preference shall be given to local (and, within that category, *Bumiputra*) suppliers, provided they meet industry standards. Petronas has tended to give wide latitude to the foreign petroleum companies in judging which suppliers meet the required quality standards, and the companies have shown a readiness to favour local sources, provided the standard is equal or only marginally lower. Most locally provided inputs have been in the nature of construction, food and security services, and so on. Up to 1982 Esso imported at least 95 per cent of its capital goods needs, and Carigali also relied heavily on imports. But the level of what is provided locally is rising. Malaysian Shipyard Engineering was providing living quarters for offshore rigs, Promet was building jackets for offshore platforms and Jurutera Consultants – a local engineering firm – in collaboration with foreign interests won a contract for the pipelines in Phase II of the natural gas utilisation scheme (Bowie, 1985).

Pakistan. Hampered by the absence of an indigenous steelmaking capability, Pakistan has relied overwhelmingly on imported equipment for the upstream part of the industry. Gas pipe of less than 6 inches in diameter used in the low pressure city distribution networks is now locally supplied under ASTM/API (American Society of Testing Materials/American Petroleum Institute) standards. With new developments in the steel producing and processing sectors, suppliers were seeking to expand the types of related material that they provide, but one problem was the reluctance of the state-owned Oil and Gas Development Corporation to credit the standards, since there is no periodic inspection procedure by ASTM/API. Most household gas appliances are assembled or manufactured in the country; the import of these items is severely restricted.

Table 9. Degree of dependence at the different stages of prospecting (in indicative percentages)

	1968		1983	
	Foreign	National	Foreign	National
1. *Geophysical surveys*				
Planning capacity	70	30	0	100
Implementation	95	5	80	20
Use of results	100	0	50	50
Inputs	100	0	80	20
2. *Photogeological and photogravimetric surveys*				
Planning capacity	80	20	0	100
Implementation	20	80	60	40
Use of results	60	40	30	70
Inputs	95	5	95	5
3. *Drilling*				
Planning capacity	60	40	0	100
Implementation	30	70	30	70
Inputs	85	15	75	25
4. *Logging*				
Planning capacity	50	50	0	100
Implementation	100	0	100	0
Use of results	60	40	10	90
Inputs	100	0	95	5
5. *Completion*				
Planning capacity	60	40	0	100
Implementation	20	80	0	100
Use of methods	60	40	10	90
Inputs	90	10	85	15
6. *Production* [1]				
Planning capacity	60	40	0	100
Implementation	30	70	0	100
Use of methods	60	40	10	90
Inputs	90	10	90	10

[1] Excluding secondary recovery, where technological dependence is higher.
Source: See the UNCTAD secretariat study: "Technological problems in the energy field in the developing countries: Development of the technological capability of the petroleum industry in Peru", as reproduced in Mayorga, A., 1985.

Peru. An assessment was carried out of the change between 1968, when Petroperú was created, and 1983 in the Peruvian (as compared with the foreign or imported) share of the various phases of the exploration and production chain. In the absence of reliable hard figures, it is based on the subjective assessment of experts working in the industry. The results (see table 9) show the relatively greater progress in developing human resources, petroleum-related expertise, than in developing local sources of material inputs that could displace imported equipment (see the last line of each item, "Inputs"). In this latter area, the change was quite limited, reflecting in large part the difficulties associated with the country's limited market size and degree of industrialisation. The main areas where local inputs are used are explosives for seismic exploration, drilling muds, auxiliary drilling equipment, pumps and small diameter pipe (plus communications, transport and construction). About 25 per cent of the cost of an oil well in the Amazonian region is spent on Peruvian supplies and services. One notable achievement has been in the construction of specialised offshore platforms, for use in the relatively calm and shallow waters off Talara. Using Brown and Root designs, a co-operative of Talara engineers and workers were producing five or six drilling platforms a year, under Belco supervision (Mayorga A., 1985).

Forward-linked: Refining and petrochemicals

Only in recent years has the presence of petroleum production in a country been considered a significant reason for developing refining and petrochemical capacity there. Most downstream capacity now present in the world was sited according to other criteria – notably the proximity of sizeable markets to which the output was thought to have more or less assured access, the availability of capital, and the industrial competence required to build and maintain complex large-scale installations. During the decade or so of perceived petroleum shortages – from the early 1970s to the early 1980s – siting decisions were influenced by new considerations. In particular, some of the major exporters then began to make claims, which were difficult to resist, to have new facilities located on their territory. They had a strong perceived need, based on the wish to diversify their economies away from undue reliance on the petroleum upstream sector and to capture more of the value added associated with their national patrimony; they had ample supplies of capital available; and they had bargaining power – the ability to grant or deny access to their hydrocarbons – with which to attract industrial partners and gain access to international markets. The export refineries and petrochemical installations now coming on stream in the Gulf and especially in Saudi Arabia are the prime instance of this, but there were other examples in Indonesia, Mexico and elsewhere. The ability of producers to price their feedstock competitively was thought to give them a very strong advantage in bidding for future downstream capacity.

The attractiveness of this option for the NODCs in 1985 is reduced on two scores, compared with the case just set out. First, with the partial exception of Mexico, this group was already much less well endowed with petroleum, capital and bargaining power than the capital surplus OPEC countries. Second, the passage of time has made downstream investment much less attractive in general because of existing and prospective excess capacity, protectionist threats to market access for refined petroleum and petrochemical products, and reduced availability of capital.

In these circumstances, the attractiveness of different options for downstream development depends heavily on particular national situations.

Countries with relatively large internal markets and with significant industrial capabilities – for example, Brazil, China and India – may find advantages in pursuing relatively ambitious programmes to cover their internal needs in these sectors. Countries with natural gas deposits for which alternative markets are limited may be persuaded of the value of developing gas-based fertiliser, methanol, or petrochemical plants, especially if the output can be absorbed in their own economy, displacing imports or increasing the inputs otherwise available. NODCs where neither of these conditions applies will not necessarily refrain from all downstream activity. Most of the countries in our sample produce in their own refineries a substantial share of their requirements for petroleum products. For reasons of national autonomy and foreign exchange savings, they may maintain or enlarge that share in the future, whatever may be the diseconomies of scale in operating refineries scaled to their national markets. (In China there has been debate about the advisability and feasibility of exporting its petroleum surplus in the form of products rather than crude oil. Hills, 1985, p. 42.) But they may be more impressed than would have been the case a few years ago with some of the arguments against downstream investments: for example, that the capital intensity of downstream projects may further skew an economy that is already allocating a major share of its scarce capital resources to the petroleum upstream; that external finance may be less available for the downstream than the upstream; and that in determining the economic viability of downstream, and especially petrochemical, projects the ability to price feedstock at attractive levels – that is, the alleged advantage of petroleum producers over non-producers – is only one, and not necessarily the most important factor. (Generally, the further downstream the product, the smaller effect feedstock prices have. For early derivatives such as ethylene, the cost effect can be substantial. See UNIDO, 1978.) For small countries international schemes – joint ventures, common markets or other international trade schemes for refined products, or simple reliance on international product markets – may seem increasingly attractive as alternatives to self-reliance in the refining sector and in petrochemicals.

The recent evidence on these questions from our sample of countries is mixed, as this selective review of downstream developments shows:

In *Malaysia* a 30,000 b/d Petronas refinery at Kerteh in Trengganu came onstream in March 1983, bringing the country's total capacity to 207,000 b/d, roughly equal to current consumption. (However, actual 1984 output was only 156,000 b/d; the balance of requirement was imported.) Authorisation has been given to expand the Esso refinery at Port Dixon by about 15,000 b/d. However, a major refinery (120,000 b/d) planned by Petronas at Malacca to process Kuwaiti crude was put in abeyance, because of the slack in the industry and the spare capacity next door in Singapore (Bowie, 1985).

A number of industrial projects designed to use the natural gas from Sabah, Sarawak and Trengganu fields are in various stages of realisation. Those under construction include a fertiliser plant in Sarawak, partly financed by ASEAN but intended mainly for Malaysian markets; methanol and sponge iron plants in Sabah, based on non-associated offshore gas; and the first phase of the Peninsular Gas Utilisation Project, which is designed to use the offshore Trengganu gas for various internal purposes (including in the first phase power generation, LPG extraction and a steel mill) and ultimately for export by pipeline to Singapore.

In *Ecuador* total refinery capacity approximately equals total petroleum products consumption, but the output is not matched to demand either in terms of products or geography. The largest refinery, at Esmeraldas, has long operated well below capacity, although since 1983 utilisation rates have improved. In that year a "Hydrocarbon Industrialisation Programme" was authorised which was to correct these deficiencies through the construction of two new refineries and the enlargement and modernisation of Esmeraldas. Joined to the refinery plans were several petrochemical projects, including propylene, urea and ammonia plants. It appeared as of mid-1985 that the Esmeraldas enlargement (56,000 b/d to 90,000 b/d) and the new jungle refinery (10,000 b/d) will proceed but that plans for a new refinery at Atahualpa and for the petrochemical plants have been indefinitely delayed (Marchan, C., 1985; World Bank, 1984a; Spurrier, 1984).

Brazil has a large refining sector based largely on equipment that is domestically designed and produced. The configuration is not well fitted to requirements, as the fuel alcohol programme has reduced demand for gasoline while new supplies of hydropower have cut the volume of fuel oil that is needed. As a result, if refinery runs are maintained at volumes sufficient to meet the requirements for middle distillates (diesel, etc.), there are surpluses of heavy and light ends which must be exported, often on terms that are commercially unfavourable. The country also has a large petrochemical sector, which was initially based largely on imported feedstock; after 1973, this caused considerable balance-of-payments problems. The increased availability of locally produced crude should ease its problems somewhat (Gall, 1985; ILO, 1982).

The refining sector of *India* has a capacity of 37.8 million tonnes per annum (1984), and in recent years has operated at about 90 per cent of capacity, the highest among major process industries in India. Net petroleum product imports amounted in 1983/84 to 2.7 million tonnes. By the end of this decade capacity is planned to expand to 61.75 mtpa. The petrochemical sector in 1983 had an installed capacity of 1.44 mtpa and a production of 1.09. In the fertiliser segment of the industry the capacity utilisation in 1983-84 was 67 per cent. The low rates of capacity utilisation, stemming from various difficulties, including problems of maintenance and supply, have been a recurring problem. Considerable expansion in this sector has been and is being made as a result of the gas reserves recently discovered, especially those of Bombay High (Tata Energy Research Institute, 1985).

As noted in Chapter 3, *Trinidad and Tobago* has also attempted to develop a heavy industry sector fuelled by its major reserves of natural gas. As of 1984, the country was the world's second largest exporter of ammonia, after the USSR, and was contemplating investments in additional capacity that could make it the largest, despite the fact that some of the existing projects were operating at a loss. It was also committed to expanding its methanol production capacity. An earlier methanol project at Point Lisas operated profitably in its first full year of commercial production (*European Chemical News*, 1985; *Financial Times*, 1985;

Economist Intelligence Unit, 1984 and 1985); however, such positive results tended to require a very low valuation (net back) of the natural gas feedstock.

In Malaysia, Thailand, Trinidad and Tobago and other countries with ample gas reserves but limited domestic markets for such products as nitrogenous fertilisers and methanol the industrial policy choices can be difficult. Heavy capital costs and current depressed markets tend to discourage new investment projects in this sector. But the future is uncertain, market balance may replace glut, and the pressure is strong for social and employment reasons to carry forward projects that offer the hope of industrial progress based on available natural resources.

Local, regional and infrastructural impacts

The relation between upstream petroleum activity and the development of local infrastructure runs in two directions. On the one hand, infrastructural development is an inducement to petroleum investment: oil companies often argue that, other conditions being equal, they are more likely to explore in countries where basic infrastucture – such as ports, roads, communications facilities, housing, water supply, health and education services – is already available (National Petroleum Council, 1982). On the other hand, petroleum development can be the spur needed for the introduction (whether by governments, operating companies or some combination of the two) of some of these facilities in regions where they were not previously available.

In serving that role, it can have a significant effect on the economic geography of a country as the physical and social infrastructure needed to support the petroleum activity contributes to changing existing patterns of economic activity. This effect is most immediately apparent in the localities and regions where the petroleum activity is concentrated. Two patterns of effects can be discerned, those which tend to be present irrespective of the terrain where the deposits and petroleum installations are located, and those which are associated with the specific kinds of remote areas where deposits tend to occur: offshore, deserts and jungles. (All of those are discussed below. For the application of these considerations to the arctic environment, see Khitoun, 1985.) In almost all the countries described in Chapter 3 most producing areas are in these special environments or in other remote areas. Since social planners are at the mercy of these geological realities, there are real limits to what can be done to integrate oil and gas operations into the wider economy and society. Downstream facilities offer somewhat more flexibility as regards siting, but factors of economic geography – the location of ports, the nearest landfall to offshore fields, and so on – can be nearly as constraining as geology.

The *general* effects are mainly those associated with the growth of "boom towns". Many of these effects are associated with large-scale industrial projects in general, and are not specific to the energy or petroleum sectors. During the development phase, there is a sharp increase in working population – demographically skewed towards young, single males – which puts the physical (housing, water and sewage, roads) and the social (schools, health) infrastructure under strain

(Yamaguchi and Kuczek, 1984). The local population tends to think that the petroleum is in some way "theirs", by virtue of its location, and that a major part of the associated benefits should accrue to them. Instead, unless the situation is properly managed, they may find that the benefits go mainly to others, as jobs are filled mainly by outsiders who spend only a small share of their income in the local economy, while they inherit most of the problems. These may include a heavier tax burden to pay for the infrastucture, a higher rate of inflation, reduced access to farmland because of speculation in land values, and the social disruption brought about by the incursion in large relative numbers of outsiders not constrained by the social mores of the local culture. The problems of adjustment may have to be repeated in reverse if the development phase is of short duration and is followed by a reduction in the required workforce and an out-migration of the redundant workers.

The difference between expectations and reality may thus be very great and may form the basis for a sectional political issue. Many countries have experienced these problems. In the Sudan a dispute over the oil area's share of the benefits of production became linked to the ongoing North-South conflict, leading to attacks on the oil camps and interruption in the development of the fields (Lesch, 1985). In Colombia three employees of an oil pipeline contractor were kidnapped by a guerrilla group, and were freed only after undertaking, among other things, to pay royalties to the local villages (Economist Intelligence Unit, 1985). The most fully developed example is probably that of the Ecuadorean Oriente, described in Chapter 3, which led to a province-wide protest action.

These are the negative possibilities. The positive ones are that if carefully planned and executed the development can lay the basis for long-term improvement in the locality's standard of living, on the basis of an improved infrastructure and enlarged employment opportunities, both facilitated by the petroleum facilities. One of the keys to this happy outcome is to moderate the rate of change – another dimension of depletion policy – and give some predictability to the course of local development. Since the petroleum development is apt to be at the mercy of geological unknowns and unforeseeable market changes, that is not easy. In many if not most countries, the requirements for large-scale and long-term planning are also apt to put the machinery of local government under great strain. One of the most interesting attempts to cope with these problems in a thorough and careful way is that of Thailand's Eastern Seaboard Project, already described, but it is too early to judge the outcome of that development. Another factor that can contribute to success is a financial mechanism that automatically allocates a share of the petroleum-generated revenues to the impacted areas. Such schemes are found, for example, in Ecuador, Malaysia and Peru, and will be discussed further below. Another policy aimed at securing the benefits for local areas is preferential hiring for local labour. This is applied, for example, in India, where "local" is interpreted to mean "from the same state".

In some instances authorities may be able to choose whether they wish to *isolate* the petroleum development from the surrounding area and population, or to promote its *integration* into the local economy and society. Both may be appropriate in particular cultural and geographic circumstances, but it is preferable to follow one of these strategies consistently rather than to follow a random and unplanned approach in which the two will be mixed. One form which this question may take is the choice between an "oil camp" (temporary settlement) and an "oil town" (permanent community) in the vicinity of producing fields located in remote areas. The availability of modern transportation has come to limit the ability of governments and companies to exercise this choice. If the oil workers prefer strongly to work in such an area on a rotational basis, flown in and out by airplane or helicopter, rather than be resettled there with their families – and there is no overwhelming cost penalty for doing so – it will be difficult to fulfil the conditions for a permanent settlement. In these, perhaps increasingly frequent, instances, isolation may be the only viable approach.

The potential of a given oilfield to support extensive growth in human settlements and in ancillary economic activity depends in part, of course, on its size and the length of its productive life. Daqing, *China*, has grown from a virtual wilderness to a city of over three-quarters of a million people over the past 30 years, on the strength of its role as the country's largest known deposit and producing field. At the same time, this concentration of effort in the already industrialised north-east of the country left the reportedly promising basins in the far west largely unexplored and lacking in the basic infrastructure needed to mount an exploration effort (Hills, 1985, pp. 19 and 56).

Examples of the special environments, and the features associated with them, are discussed in the following sections.

Desert

In *Egypt* most production and reserves are either offshore (Gulf of Suez and Red Sea) or in desert areas (Sinai and Western Desert). Even the main support bases for offshore operations are located in desert areas. Two of these on the coast of the Gulf of Suez, Ras Shukheir and Ras Gharib, offer an interesting contrast in how such bases may be organised. Ras Gharib was established before the air age, in 1936, to support nearby onshore fields although it has come to be used also as a base for offshore fields that have since been developed; it is organised as a permanent settlement, an "oil town", with a population of 50,000. Workers live there with their families, and such accoutrements of fixed abode as hospitals and schools are also present. Ras Shukheir was established in the mid-1960s specifically as a base camp for offshore operations. By that time the existence of air transport offered the option of flying workers in and out on regular shifts, avoiding the necessity of relocating their families. Ras Shukheir was therefore built as a camp. It accommodates some 1,500 people, all male, who operate on a cycle eight days work, seven days off. Daily flights take the workers going off duty to Cairo for their free time, and bring them back a week later. Facilities are restricted mainly to those needed for such an existence: lodging, food service, recreation and health care (Norconsult, 1983d).

In contrast to these areas on the Gulf of Suez, where petroleum activities have generated little in the way of viable autonomous activity, the oil areas in the *Western Desert* are the scene of ambitious plans for regional development.

Government plans envisage the settlement of 750,000 people in the area by the year 2000. The oil companies' contribution is expected to take the form of finding and producing underground water supplies adequate to support the new towns and irrigated farms that are planned, and assisting in the construction of roads; the General Petroleum Company planned to build as much as 2,300 km of roads in the Western Desert.

Experiences elsewhere also fit the pattern of the old and new styles, as described for Egypt's Gulf of Suez area. In Talara, on the coastal desert in northern *Peru*, an oil town has grown up over the course of the past century; it can be compared to Ras Gharib as a permanent settlement based mainly on petroleum workers and their families. (Jobs in the industry are in fact often passed from father to son, provided the son is qualified, in what is one of the world's oldest continuously producing oil provinces.)

The largest producing field of *Tunisia*, El Borma, is located in the Sahara, in the extreme south of the country, and is remote from any permanent settlements. It was discovered and came into production in the 1960s, and a similar logic was applied as in the case of Ras Shukheir. There is a camp which accommodates 500 workers, some 300 of which can be found there at any given time. They are flown in and out on a cycle of three weeks' work, two weeks off, three weeks' work, one week off. Although in the lower skills category many workers come from other parts of southern Tunisia ("local" workers are given preference, but the term is relative, the nearest town being 160 km away); almost all of the professionals live in Tunis. Some infrastructure that has been built for purposes of petroleum development in the desert – for example, tracks permitting motorised access to El Borma or to exploration sites elsewhere in the Sahara, or water wells that were drilled initially for purposes of exploration or production – may have some limited utility for non-petroleum purposes (see Schliephake, 1977).

Jungle

The contrast between old and new styles also characterises jungle settlements, to some extent. Where oil production in such an area began before the air age, as in a remote part of Assam in north-eastern *India* a century ago, there was no alternative to the construction of a permanent infrastructure in the area; there the company had to assume responsibility for developing and to some extent operating transport links, hospitals, schools, and so on.

In *Peru* oil production from the jungle is more recent, but it is located near a settlement that owes its origins to an earlier commodity boom, that of rubber in the early years of this century. This city, Iquitos, offered a natural supply and service base when oil exploration began in Peru's northern jungle in 1970 and when production began, with the opening of the North Peruvian pipeline in 1976. It now has a population of 100,000. This makes it more of a permanent settlement, with more possibilities for a resident workforce with families, than is the case, for example, in the jungle producing areas of Ecuador. Around 900 Petroperú employees are based in Iquitos. A small refinery is located there. None the less reliance on air transport remains substantial, both for communication between

Iquitos and the surrounding camps and for links to Peruvian cities in the Sierra and coastal regions. No road has been built or is envisaged between Iquitos and the western part of the country (the Atlantic can be reached by river, via 4,000 km of the Amazon network), given the long distance and difficult terrain involved (Norconsult, 1983c).

In *Ecuador* a road has been built linking the established road network to Nueva Loja, the principal town in the jungle producing area, and beyond that to some of the principal oil camps, such as that at Shushufindi. Altogether, the province's road network now includes about 440 km of principal and 330 km of secondary roads and 130 km of tracks. Almost all of this has been built since oil production began in the early 1970s. In addition, there are now 22 secondary and tertiary airports with asphalt runways. These improvements to the physical infrastructure may help account for the fact that there has been a rapid influx of population into the province since oil production began, including many agricultural settlers as well as support workers for the petroleum activities. Between 1962 and 1982 the population of Napo Province nearly quintupled, to 113,000. Other aspects of the infrastructure, however, have lagged behind the transport network. Drinking water, sewage and telephone service are still inadequate, as are schooling and health services. The companies provide well for their own employees in the camps, but the general population is less well served. Informally, and on occasion, the company health services have been made available to non-employee local inhabitants. It is dissatisfaction with these problems of infrastructure and support services that led to the region-wide protest that has already been referred to in Chapter 3 (Marchan, 1985).

In *Guatemala* exploitation of the jungle fields of Rubelsanto, Chinaja and Caribe led to the construction of all-weather roads linking those areas to Guatemala City and to the Atlantic port of Piedras Negras. The latter spur parallels a pipeline built with excess capacity so that with the addition of further pumping capacity it could carry up to 50,000 b/d of crude (1985 national output about 3,000 b/d). Some subsistence farmers have settled along these roads near the camps, and a small school and clinic for their use have been established by the operator, Hispanoil.

In East *Malaysia* a similar population explosion and associated development have taken place in Bintulu, in Sarawak, which became the base for the offshore activity that began there in the 1960s. It grew from a sparsely populated coastal settlement to a town of 14,700 in 1977, 28,000 in 1980 and 36,000 in 1985. Large natural gas finds led to plans to build a liquefaction plant and associated processing and port developments at Bintulu. Huge capital investments were involved: M$7.2 billion for the LNG project, M$560 million for the port, and M$737 million for the urea/ammonia plant. Because of the scale and importance of the project and the potential for disrupting a region where there were special cultural and social sensitivities (the "native races" of Sarawak are classified as *Bumiputra* and assigned special rights in the Malaysian Constitution), great care went into the planning and execution of the project. An independent Bintulu Development Authority was set up, and a master plan was drawn up with the help of international consultants covering not only the industrial installations but also the related town planning. Execution took place over seven years, and in 1983 the LNG plant came onstream only two weeks late. The successful urban achievement included the construction of a satellite town with 1,650 housing units and all the associated amenities, and a sensitively planned expansion of the Malay village part of the old Bintulu Town. The latter involved full participation by the Malay kampong people and resulted in a style of development suitable to extended families and minimally disruptive to the existing mode of life. The Bintulu experience is perhaps one of Malaysia's most successful examples of long-range planning. A less creditable fact is that over 15 years after production offshore Sarawak began, there still is no satisfactory north-south road linking the main population centres of the state.

Offshore

If petroleum resources are located offshore, that gives a special cast to most of the issues that are reviewed in this chapter – employment, industrial and local/infrastructural effects. Given the more sophisticated technology employed, the jobs created by offshore activities are more likely than onshore projects to require skills and experience that are not locally available; hence recourse to expatriates and non-local nationals is likely to be greater. Backward-linked industrial activities are similarly more demanding and capital-intensive, and so reliance upon imported equipment will probably be higher – as we have already seen, for example, in the cases of Brazil, China and India. The kinds of environmental and safety issues that arise are somewhat different (see Chapter 6). And the local and regional impacts take the particular form of creating or adapting coastal installations to provide maritime supply and support services and land-based gathering, processing and entrepôt points (Cairns and Rogers, 1981; Gilbert, 1982, 1983).

The problem of phased and temporary activity (peaking during the development phase, with associated "boom-bust" problems) is equally true of offshore as of onshore petroleum projects. The lead times involved tend to be longer. This period of grace can be advantageously utilised as a planning period for the urban and infrastructural development associated with the petroleum installations. But the uncertainties inherent in petroleum production make the process more improvisational and less neat than might be desired. Even after production has begun, surprises are still common regarding the recoverable potential of a field; moreover the results of exploration in adjoining areas will probably not be fully known for some time.

If the support bases for offshore activity are not wholly dependent on that activity, it is easier to adapt flexibly to the vagaries of the exploration and production process. Much of the support activity for India's Bombay High fields has been based in the port of Bombay, which has been expanded and modernised to accommodate the ONGC's needs. As it still remains congested, however, the Government is considering the construction of a new port south of Bombay that would be largely dedicated to ONGC requirements. The approach is incremental and pragmatic. In *Brazil* the development of the Campos fields – the most significant concentration of reserves and production in the country – has been carried out mainly from the base of Macaé, a small port located 190 km north of Rio de Janeiro with a varied history as a slave-trading centre, an entrepôt for farm products, a railroad-canal transfer point, a fishing port and a weekend recreation site. Its population has grown from 12,000 in 1950 to 55,000 in 1980 and perhaps 100,000 in 1985. Despite some "boom-town" type problems – housing supply lagging behind demand, accelerated inflation and attraction of a floating population in search of work – it has adapted reasonably well. In southern *China* several localities appear to be competing as supply bases for the offshore work. However, as it remains far from clear how far this exploration activity will be converted into development and production, there is a risk of the supply outrunning the demand for land-based services. In the case of one of the ports, Zhuhai, there appears to be hope that petroleum activity could assist its development into a significant new general-purpose port serving the rapidly developing Pearl River Delta and Southern Guangdong Province, but to the extent that the financing for this growth depends on foreign sources it may be hostage to the outcome of the petroleum search.

Regional and other special purpose funds

Several countries have special financial mechanisms designed to plough back to their petroleum producing regions a part of the associated revenues, usually for the purpose of covering the costs of extra infrastructure required by the petroleum impact and sometimes for other, more general reasons. Descriptions of regional funds in Ecuador and in Peru, and the contentious issues associated with them, have been made in Chapter 3. In *Malaysia* the production-sharing contracts provide that the contracting companies shall pay directly to the state governments concerned 5 per cent of the petroleum revenues that are due to the authorities. In 1985 the annual sums involved are as follows: Trengganu, M$235 million; Sabah, M$90-100 million; Sarawak, M$120-130 million. The states are free to spend the funds as they wish, and appear to regard the share as a reasonable one. Much local development, especially in Trengganu, has been financed with these funds. In that state, electricity supply, health services, telephone connections, water supply and other public amenities increased substantially in recent years, as these funds became available, but they nevertheless have tended to lag behind demand, which has been expanding even more rapidly as petroleum development goes forward. In *Brazil* states and territories receive 4 per cent of the value of the petroleum extracted from their territories; 1 per cent goes to the municipalities. Prior to 1957, 5 per cent was paid to the states, which were supposed to pass on one-fifth of that to the municipalities, but there were complaints that the latter did not always receive their share. Priority in spending the funds was to be given to infrastructure, especially road-building and electricity. These funds paid by Petrobrás have provided significant shares of the revenues of certain states, e.g. Bahia, Sergipe and Alagoas, in the relatively poor north-east, and of certain municipalities (Randall, forthcoming).

Several countries have established financial mechanisms aimed at furthering other, more general purposes. In *Gabon* the contract governing the activities of the local Elf affiliate stipulated that the company should set aside each year funds for investment in projects likely to further the country's economic development. This "provision for diversified investments" was to be put into petroleum-related or other areas. The funds were to be invested within five years of being set aside, and 65 per cent of the equity in the resulting ventures transferred to the Gabonese State. Investments have been made in the fields of cement, cellulose, shipping, hotels, paint and computer services (*Elf Gabon*, 1979). *Guatemala*'s basic petroleum legislation (1983) laid down, in advance of large hydrocarbons discoveries, provisions governing the expenditure of revenues from future petroleum production. Such revenues were to be paid into a Fund for National Economic Development, whose purposes were to be exclusively the development of the interior of the country and the development of new and renewable sources of energy (*Ley de Hidrocarburos y su reglamentación*, Guatemala, C.A.: Ministerio de Energía y Minas, 1984). In 1985, as exploration activities were accelerating, the statute establishing this fund was being drafted. In *Malaysia* a Trust Fund was being established in 1985 to which Petronas was contributing M$200 million in the first year, with increased contributions in the future, which is to be invested against the day when petroleum revenues decline. States and private companies were being encouraged to contribute also. The funds were to be invested in blue chip stocks, both in Malaysia and abroad (*New Straits Times*, 1985). As noted in Chapter 3, both *Trinidad and Tobago* and *Ecuador* have made use of special funds for macro-economic purposes related to their petroleum revenues. In the former case, it was to earmark and set aside the funds so designated until the industrial development projects in which they were to be invested became ripe. In the latter, the various accounts assigned automatically a certain share of the petroleum proceeds for specifically designated purposes. This carried the advantage that the dangers associated with discretionary spending were reduced, but the drawback that budgetary flexibility was diminished and expenditure control in the earmarked areas was made more difficult (World Bank, 1983a and 1984b).

Chapter 6

Social issues

Causality is not easily established with respect to the various social effects, both positive and negative, that often accompany, and are widely believed to be caused by, petroleum activities. On the one hand, where public sector programmes for social improvement are paid for in whole or part by revenues generated by petroleum production, it may be argued that resulting changes in the indicators "resulted from" the petroleum production. In Chapter 3 we saw how dramatic improvement occurred in several social indicators during the 1970s in Ecuador and how expenditures on primary education increased dramatically in Nigeria. It seems nearly certain that much of this effort and some of the resulting achievement (in health and literacy levels, for example) would not have taken place in the absence of petroleum-generated revenues, but it is impossible to know how much. On the other hand, the petroleum industry – the publicly as well as privately owned segments of it – is often blamed for various social ills, such as destruction of traditional values and ways of life, that can be noted in areas where oil and gas production takes place. Again, while it is plausible that the social change that is deplored is accentuated by the presence of petroleum operators – who tend to come from the outside and to belong to a technologically advanced culture – there is no way to judge confidently how much change would have occurred anyway through the intrusion of modernity into traditional areas via other channels, such as television, tourism, and so on.

Rather than engaging in the futile exercise of trying to allocate causality among different sources, the discussion here merely indicates the nature of some of the more prominent social issues that arise in relation to petroleum development, with examples as appropriate from our sample of countries.

It will be noted that wages and working conditions, traditional ILO concerns, do not as such figure on this list. Although included in the research design, they were not identified in the national case studies and other research done for this project as problem areas in themselves. As in other studies carried out by or for the ILO (ILO, 1977; Norconsult, 1984), a number of the country case studies observed that, in general, petroleum workers enjoyed wages and working conditions superior to those in other sectors. The capital-intensive character of the industry makes it both possible for employees to be relatively well looked after (there are not too many of them) and sensible to invest in good industrial relations (the cost of work stoppages, in terms of lost production, can be very great, especially in the upstream). Particular problems, of course, continue to exist: random examples encountered in this research, for instance, include issues of workers' safety in *Brazil*; of pay and conditions of service in firms being nationalised in *Bangladesh*;

and of the right of local unions to represent workers on offshore installations in the *Côte d'Ivoire*. In some countries industrial relations issues are complicated – to the benefit of neither workers, employers nor governments – by the existence of a multiplicity of trade unions. This is caused partly by the wide range of different skills and trades required in the sector; sometimes employer or government resistance to the organising of the workers is a contributing factor. But in general, securing for petroleum workers a fair share of the benefits of petroleum production does not appear to be a major problem.

In a few countries with powerful petroleum workers' organisations, issues of a quite different character have arisen. These concern the question whether petroleum workers may be getting a disproportionate share of the benefits, to the detriment of society at large. One issue concerns access to jobs. Where the union has a dominant voice in hiring decisions, the number of employees may be greater than required and the selection process may give undue weight to criteria (such as standing with the union) other than ability and considered policy objectives (such as favouring workers of local origin – see issue No. 3 below). Another concerns the level of remuneration, which may be higher even than conditions in the petroleum industry warrant and which in its standards of probity may not withstand scrutiny. The project research has not shown this to be a widespread problem, but in respect of a few countries it has received attention in the open literature (e.g. Prevot-Schapira, 1982).

Impact of expanded commercial energy consumption

It is appropriate to put at the head of the list the most immediate and direct social impact, that which comes from the use of petroleum products which would not otherwise have been available. The widespread use of oil products for transportation, electric power and other purposes has of course altered millions of lives; however, local production influences this process only in degree, as the same consumption patterns, based on imports, occur in many places where there is no indigenous oil production. By contrast, in the case of natural gas, local availability can have a substantial impact on the energy use patterns of the final consumer. In a number of countries which have indigenous natural gas deposits *and* patterns of population density sufficient to render economic the construction of residential distribution networks, the use of gas for domestic purposes has become widespread. Its qualities of cleanliness and convenience are affecting for the better the quality of many lives. *Pakistan* is one of those countries. There, gas consumption in the residential sector grew from 1969 to 1984 by 25 per cent per annum. This growth was reflected in an increase in the number of households served by the network to over 900,000 by June 1983. Many of these previously relied on non-commercial fuels, such as firewood or dung. *Bangladesh* and *Egypt* are other countries where the domestic use of natural gas has been growing rapidly. *India* may soon follow, as work proceeds on the South Bassein and other offshore gas fields.

An associated benefit of natural gas development is often the greater availability of liquefied petroleum gases ("LPGs" – butane and propane), which in some natural gas deposits are present in concentrations that justify the construction of separation plants. (Oil refineries also yield some LPGs, usually in smaller quantities.) Because LPGs are usually distributed in steel cylinders weighing (in one common version) only 14 kilograms, their use does not require the massive and lumpy investments needed for natural gas pipeline networks; they can therefore potentially reach better into areas of lower population density. In *Bangladesh*, *India* and *Thailand* it is government policy to expand the use of LPGs for domestic purposes, a possibility which the growing availability of local natural gas makes possible, in order to ease the firewood and fuel problems, which have been becoming increasingly difficult. In *Mexico* it has also been found that the firewood shortage which exists in certain areas could be eased by greater use of LPGs, with added benefits in terms of convenience and time spent on wood-gathering (Evans, 1984). Even though the weight and cost of the LPG cylinders tend to restrict their use in the first instance to cities and towns, rather than rural areas, the resulting reduction in the urban demand for firewood can – by reducing the competition for available supplies – indirectly ease the fuel problems of village-dwellers and other rural people as well. Moreover, innovations in cylinder design, size, distribution and methods of finance are making it feasible to get LPGs delivered – provided supplies are adequate – to progressively smaller and more dispersed settlements. Since in many parts of the world wood-gathering is women's work, a greater use of LPGs in those areas would bring attendant advantages for them and the children they care for.

Safety and environmental concerns

Over the years, the scope of these concerns has steadily broadened, so that they now include not only safety at the workplace, but also pollution and other direct environmental problems, and even indirect and long-range effects such as those associated with "acid rain". Petroleum industry operators have several good reasons to aim at a high standard of performance: the costliness of replacing damaged equipment and of production shutdowns caused by accidents; legal requirements; and the need to protect their reputations in a world where publicity travels fast. Despite occasional oil spills and other dramatic incidents, the safety and environmental problems involved in petroleum production, processing and distribution are relatively localised and moderate compared with other major extractive and processing industries and are minor compared with the impact of petroleum-utilising activities such as motor transport and power generation. (For a general review of issues concerning safety, health and the working environment in the industry, see ILO, 1986, II.)

All the same, there are dangers associated with most stages of the chain from exploration to use. At the exploration stage these are relatively minor, and may involve the cutting of trees for conducting geophysical surveys and the damage or nuisance caused by explosive charges. In the drilling and production phases there is a problem of disposal of drilling muds and of saline water extracted with the oil or injected to assist recovery, and a risk – much reduced nowadays but still real – of blowouts. Offshore production involves special risks for divers, platform workers and for those involved in marine and helicopter supply operations (ILO, 1978). Transportation and distribution may involve leaks, spillages and fires – which may be minor if from a pipeline or major if from a large tanker. Refining requires a supply of fresh water, and produces effluents that should be treated before disposal.

The more widespread use of natural gas and LPGs referred to above involves what are in some respects problems of a new order, as these fuels become more widely used in households, many of which will be unaccustomed to the technology and risks involved. Legislation, technical standards and inspection procedures need to be developed covering the production, processing and distribution phases and the appliances used in final consumption. The explosion with large loss of life that occurred at a natural gas processing plant in Mexico in 1984 exemplifies the risks that exist at the more centralised points of the system. It is becoming more widely recognised that the more dangerous installations, such as LPG bottling plants, should be sited outside major cities. Problems nearer the point of consumption – residences and small commercial enterprises, for example – receive less publicity, but may be collectively more important. In Pakistan, for example, it is reported that thanks in part to co-operation from management and unions the safety situation at oil and gas drilling and production sites, and at large gas utilising factories like cement, power and fertiliser plants, is relatively satisfactory. Total occupational injuries per 1,000 workers per year ranged from 40 to 43 during the years 1981-83 (minor + serious + fatal). However, although data are inadequate the situation is thought to be less satisfactory in the case of smaller gas consumers.

In some NODCs as elsewhere, impact assessment procedures are used to promote the timely and comprehensive planning needed to minimise the negative environmental impacts of petroleum and other large-scale industrial projects. Brazil, Malaysia and Thailand are among the countries which provide for the use of such an approach. But what matters is not what it is called but how it is applied. The phrase "environmental impact assessment" (often in capital letters) sometimes carries a mystique, implying the use of sophisticated techniques, that is not warranted. So used, the phrase can arouse unjustified fears among industrialists that it is only a means of delay and obstruction and unjustified hopes among planners that it offers an automatic solution to complicated planning problems. In fact, if successful the approach is not a substitute for the usual ingredients of judgement and dialogue between authorities, enterprise management, workers and other interested groups but another name for exercising them in a timely and organised way.

Local vs. migrant labour

The reference is not to the subject of nationals replacing expatriates at the upper skill levels – which of course is an issue, as discussed in Chapter 5 – but rather of *which* nationals benefit from petroleum-related jobs, those indigenous to the producing areas or those who move in with the rigs. It is commonly the case that *(a)* residents of petroleum-producing areas resent the capturing of jobs in the industry by those from outside the area; and *(b)* the local labour force in such areas is ill-suited for the requirements of the petroleum industry. The involvement or non-involvement of local residents in planning the developments surrounding the petroleum installations may also be an issue. In *Thailand* a survey of workers at the Sirikit field in the central part of the country found that most of the professionals were expatriates, most of the skilled labourers came from Bangkok, and most of the total labour force from other parts of the country; only about one-seventh of the workers came from the district (Lan Krabu), and they almost all held unskilled jobs (Chulalongkorn, 1985). In the area where the gas-based Eastern Seaboard project is located, expectations of jobs among local applicants have been high, well ahead of the creation of jobs. A registry was established and restricted to applicants who can prove local residence, and the provincial technical training school has initiated a course to train skilled workers for the incoming industries. But actual hiring of locals has been slow to come (Tingsabadh, 1985). We have already seen that the problem exists in the *Peruvian Oriente* and in *India*, where policy requires preferential treatment for state residents, as well as for weaker sections of the community designated as "Scheduled Castes and Tribes". The Sui gas field in *Pakistan* is located in a less developed area (Bugti land), where the offer of employment to local residents was only hesitantly made by the producing company, was then rejected, but ultimately was realised on a limited basis.

Income distribution effects

The most powerful feature of petroleum-generated revenues from the standpoint of their influence on income distribution is their tendency to be concentrated initially in relatively few accounts, especially those belonging to state oil companies and government entities. Everything then depends on decisions made by a relatively small number of persons, and few decisions are neutral in distributional terms. Price subsidies for petroleum products may benefit relatively large numbers of people of various social classes, but tend to favour urban over rural residents. Subsidised natural gas prices are especially prone to benefit a minority of middle- to upper-income consumers. Reductions in the tax burden may have favourable distributional effects if the initial tax incidence is relatively broadly based. Improvements in remuneration can be granted to those employed by the government, which will tend to strengthen the urban bias that may already be present in subsidy programmes.

The revenues may also be used to reward those engaged in petroleum production. Particular care is needed in establishing remuneration policies for petroleum workers. In addition to the labour market difficulties that may result from inappropriate policies, already discussed in Chapter 5, preferential treatment for petroleum workers may become a social and political issue, as it has for example in Mexico and Trinidad and Tobago (Grayson, 1981; Prevot-Schapira, 1982; Norconsult, 1983c). While there is no magic way to avert these dilemmas, the appropriate guide-line is to set remuneration policy in each part of the petroleum sector according to the degree of geographic and industrial mobility, and hence the applicable scope, that characterises the labour market in that segment.

Investments in infrastructure and in industrial projects are other principal ways of spending the revenues. Distributional effects depend crucially on the nature of the projects and on their economic viability or lack thereof.

The integrity and transparency of government accounts are necessary if the reality and suspicion of improper payments are to be eliminated or minimised. A similar standard of probity will need to be seen to apply to private companies and organisations of workers in the petroleum sector. Corruption is always a danger where such large amounts of money are spent in discretionary fashion. As seen in the preceding chapter, specially created accounts are sometimes used not only to isolate funds in their macro-economic effects or to allocate them for regional or other designated expenditures but to reduce the element of discretion in their disbursement, replacing it with a degree of automaticity and hence reducing the scope for misappropriation of funds.

New life-styles and old values

Several characteristics of petroleum activity combine to create the potential for a socially disruptive conflict between old and new values and life-styles: the tendency for upstream sites to be located in remote areas which are less urbanised and generally less developed; the common reliance on a workforce brought in from outside the immediate community, frequently from outside the country and of an alien culture; the cutting off of those workers from the normal social inhibitors of family and community, through a work pattern designed for men on temporary assignment; and the injection into this scene of a large increment of disposable money income. If not deliberately muted through appropriate policies and behaviour, the resulting cultural encounters can be sharp. This clash can occur in virtually all producing countries, industrialised and OPEC as well as NODC. The 1978-79 Iranian revolution is perhaps the most dramatic instance of social change in which these factors, as well as others particular to that country, played a role.

Most of our sample of NODCs are subject to this conflict to some degree. In some it takes the particular form, dealt with in the following section, of encounters with indigenous tribal peoples. In others it is a contemporary manifestation of an issue with deep historical roots. China is such a case. Although the contemporaneous experience of resident expatriates in China has been brief, and information about it is sparse, it is clear that both expatriates and Chinese are being urged to be cautious about their personal relationships with each other (Bennett, 1984). So far the separation of the foreigners into residential enclaves has limited the scope for "spiritual pollution". But if the scale of petroleum activity and hence the size of the expatriate community were to expand significantly, it is likely to be increasingly difficult to sustain this policy of isolating the source of the disturbance.

The experience of *Malaysia* in its peninsular east coast province of Trengganu is an interesting one. When petroleum development began, Trengganu was characterised by a traditional Malay way of life based predominantly on Islam.

The infrastructure was less well developed than in other parts of West Malaysia. The discovery of oil and gas resources offshore for which Trengganu is providing onshore support has brought significant social change to the state. Labour force participation rates increased for females and for the elderly (55-64), partly to compensate for the departure of young people from farms to the cities, which in turn was partly job-related; the participation rate for the 15-19 age group declined slightly, as educational opportunities rose. Nuclear families have increased in number, partly encouraged by home ownership schemes or company-provided quarters introduced to alleviate the housing shortages which were created when large numbers of foreigners and out-of-state Malaysians began to move in. The "consumer society" tends to be embraced by the young and regarded circumspectly by the elderly, who fear that traditional values and religion will be undermined. Perhaps in reaction, the resurgence of "fundamentalist" Islam which the country has witnessed in recent years has been stronger in Trengganu than in many other parts of the country.

The Government and the oil companies involved have been aware of the sensitivities and have sought to cushion them in various ways. The state's share of petroleum royalties has provided it with the means to build much of the needed infrastructure. Carefully planned and widely accepted new towns were built in Rantau Petronas and Kerteh New Town. Emphasis has been put on education, which is a concern of local residents – eager to capitalise on the employment possibilities offered by the petroleum activities – and of the company staff who have moved in from other parts of Malaysia. In part these steps have been intended as encouragement for oil company workers to relocate with families to Trengganu, although during a transitional stage they were enabled to leave their families behind and visit them at weekends or on off-duty shifts.

In sum, although problems have been inherent in the exercise, culturally sensitive management, by the authorities and oil companies involved, of the social problems associated with petroleum development has enabled the state to make a relatively successful adaptation to its status as a petroleum-producing area.

Tribal peoples

A number of remote areas into which petroleum explorers have penetrated in recent years are inhabited by indigenous tribal peoples who occupy special positions in their countries. Little exposed to contact with those outside their own groups, they are socially and biologically vulnerable, to a varying but significant degree, to the influences to which they are subject as outside agents move to explore or occupy lands which they have traditionally occupied. Almost all of the NODCs in our sample, plus others such as Bangladesh and Guatemala, have such minorities who have been affected by petroleum exploration and/or production. Very difficult situations may arise when such groups encounter petroleum operators whose right to explore has been granted by national authorities according to an agreement of which the tribal group may be unaware or to which, if it is aware, it may well be opposed. The national authorities commonly attach a higher priority to the search for petroleum than to peacekeeping with the tribal group or otherwise protecting their interests, but that does not mean that they will have reconciled the conflicting claims before the geophysical survey crews arrive on the scene. There are cases (e.g. in the Chittagong Hill Tracts in Bangladesh, Pakistani Baluchistan, southern Sudan) where geologically prospective areas lie in regions that are inhabited by tribal peoples (or other ethnic groups different from those which control the central government) and that are not, in practice, under the effective control of the central government. In some of these instances governments have none the less granted permission for exploration and/or production in these areas. In such circumstances the companies, or their subcontractors, have on occasion to act in a policy or at least administrative vacuum in which they must attempt to resolve the conflict on the ground. Controversies have arisen regarding the way in which this has been done in Brazil, Ecuador and Peru among others (see *Survival International Review* and *Survival International News*, various issues); oil exploration in Bangladesh and development work in the Sudan (as noted above) have been interrupted because of problems – associated with the ethnic conflicts – of physical security of company personnel.

Although petroleum development and other forms of natural resource exploitation often bring the problem to a head – because they tend to generate occasions for physical encounters between the tribal people and outsiders more or less backed by the central government – the issues involved go far beyond the boundaries of any one sector. Complex historical, cultural and legal questions are involved. Much of the conflict revolves around the issue of land and resource ownership, which tends to be differently understood and legitimated in the tribal cultures from that of most central governments (Swepston and Plant, 1985). Although virtually all developing countries vest title to subsurface minerals in the State, access to those minerals normally requires ownership or leasing of the surface area above. Where those who effectively occupy the area are prohibited from owning, do not own, or do not understand the significance of ownership of the land, a decision to push ahead with exploration may lead to conflict.

The dimensions of this issue are likely to continue growing. By a conservative estimate the number of tribal/indigenous people in the world exceeds 300 million. Organisations to defend their interests are becoming more numerous and professional. Legislation to protect their interests is proliferating, especially in the Americas, although the effect given to it varies considerably from one country to another. The proper context for dealing with the issue is a broad one covering land rights, settlement policy and tolerance of ethnic diversity. Close co-ordination is needed within the government by the various ministries and agencies responsible for different aspects of the question. But given the slowness and difficulty in resolving it and the inexorable trend for petroleum activity to push into the remote areas where tribal people tend to be found, there are likely to be more instances where petroleum operators trigger off incidents the fundamental causes and characteristics of which have little to do with petroleum.

Chapter 7

Summary: Policy issues

The breadth of this study and the variety of countries and circumstances that the subject encompasses make it impractical to state conclusions in the conventional manner. The exercise has been essentially to identify, categorise and clarify the developmental issues that upstream petroleum activity entails for the NODCs. By way of summary, it may therefore be appropriate to concentrate on the principal policy issues that have emerged as having broad application.

One question with which this study has not concerned itself is one which for many governments is the highest priority petroleum issue, that is how to attract upstream petroleum investment into their countries or how to expand the effort that is already being made. The concern here is with countries that are already oil and gas producers. It is about how a given level of effort can best be put to developmental purposes, not about what determines that level. It is well to be clear, however, that all the other policy issues to be reviewed below are dependent on the basic contractual and institutional framework governing the upstream effort. If this does not provide sufficient means and incentives to generate and sustain a successful exploration effort, then the other policy issues do not arise. If the requirements imposed on upstream operators with the aim of enhancing the socio-economic impact of petroleum production are so elaborate as to deter or interfere with the operators' primary mission, then of course the purpose of those requirements will have been defeated. The fact that we are here concerned with what in some respects are secondary priorities should not cause us to lose sight of the fact that there are primary ones; and they can be influenced by the way the secondary ones are handled.

This is a point that at least some governments have grasped clearly, as was revealed during the research for this study. Where it has been judged that investigation of social issues related to petroleum production could prejudice future investments in exploration, the ILO was discouraged from proceeding with those aspects of the research.

What we are reviewing here are not the policy issues internal to the petroleum sector, which are of a more specialised and technical nature, but those which affect the relationship between that sector and the broader economy and society. The form and priority of the various issues will vary considerably from country to country – according to its size, level of production and of net exports or imports, duration of petroleum experience, respective roles of the private and public sectors, etc. This will be clear from the earlier sections of this report, where most of the policy questions have implicitly if not explicitly emerged. None the less some of the principal issues have a widespread if not universal relevance.

One important subject which does not lend itself to precise measurement or analysis is the effect of government actions on public *expectations*. The discovery of oil or gas tends almost everywhere to give rise to a widespread belief that "things will be different", and easier, in the future. These expectations can be quickly translated into claims to various kinds of entitlements which, if granted, could easily surrender in advance whatever welfare gain results from the new production. And that gain itself should not be counted on too much until it is realised; instances are common of hydrocarbon deposits turning out to be much more modest than they first seem, and prices too may turn out to be lower than expected. Governments cannot completely suppress this tendency towards inflated expectations, which seems to be an intrinsic feature of petroleum experience. But they can combat it – principally through their symbolic acts, rhetoric and information policy. The message should be that the criteria for investment projects are no different with petroleum revenues (or with larger revenues) than without them, that those working in the petroleum sector must perform according to the necessarily high technical standards which characterise petroleum operations, and that those in other sectors will have no easier a task than before – in fact, their task could be harder if the greater availability of foreign exchange means greater competition from imports or if their more able personnel are drawn away into the petroleum sector.

A second point of general application is that the broader social and economic concerns dealt with in this report, even if judged of lesser importance or immediacy than the primary petroleum and revenue issues at stake, are worthy of *timely* attention by the relevant policy-makers. Who these are is less obvious than in the case of petroleum policy in a narrow sense, but it is probable that ministries responsible for economic, industrial and financial affairs, for labour and personnel questions and for relations with regional and local authorities should have significant roles to play in setting the broader policies related to petroleum exploitation. A consultative process is needed which allows not only these parts of the central government, but also petroleum industry management and workers and relevant local authorities, a legitimate opportunity to influence decisions related to their interests and competence. How this should be done in an institutional sense depends on national particularities, but as a general proposition it appears that the countries which have worked out means for incorporating these broader constituencies in their petroleum policy-making have had more satisfactory experiences than those which have taken a narrower approach dominated by a petroleum ministry and/or state oil company.

For example, Malaysia's experience has been a successful one. There, broader interests have two primary channels for influencing petroleum policy. First, the board of directors of Petronas, the state oil company, includes, ex officio, senior representatives of the Ministry of Finance, the Ministry of Trade and Industry, the Prime Minister's Department, the private sector and the oil-producing states. Second, a National Petroleum Advisory Committee, a senior inter-agency body, is responsible for giving oil and gas policy advice to the Prime Minister, who in effect serves as the minister for petroleum. There is no petroleum ministry; the energy ministry's mandate does not include oil and gas.

The following chart illustrates in a schematic way how socio-economic components can be incorporated into overall petroleum sector planning, taking advantage of the long lead times inherent in the activity to incorporate broader planning considerations and noting the appropriate agent of each activity. Although the categories are in some respects arbitrary and the chart is neither detailed nor comprehensive, it gives an overview of the process required (see table 10).

The numerous references made in the table to Environmental and Socio-Economic Assessments (EIA) and Statements (EIS) should not be taken as a recommendation for a specific methodology but as encouragement for all systematic efforts to weigh, in advance of execution, the impact of large-scale petroleum projects on the physical environment and on the communities where they are located, and to adapt the projects as appropriate in the light of those assessments. (There is a vast literature on EIAs, much of it of limited usefulness because of its abstruse or technical nature. The following sources may be useful because of their combination of breadth, balance and/or clarity: UNEP, 1980; *Environmental planning for large-scale development projects*, 1983; Carley and Bustelo, 1984.)

A third general point concerns institutions in the oil and gas sector itself. Most NODCs with upstream petroleum activity have seen fit to establish state petroleum enterprises (SPEs) as the guardians of the public interest in the sector. That in itself has not been a barrier to effective co-operation with private sector companies, which in most countries have shown an ability to work flexibly and pragmatically with SPEs (United Nations Centre for Natural Resources, Energy and Transport, 1980). A more common problem, perhaps, has been tension between the SPEs and the ministries or other government bodies to which they are nominally responsible. The decision to have an SPE does not necessarily entail any conclusions regarding the extent, if any, to which that enterprise should be engaged in operations. The precise location of the line dividing SPE from private sector operations is less important than the degree of co-operation and constructive working relations that exists between the two. Research done for this project shows that in a decentralised multinational petroleum enterprise many of the skills and other resources embodied in the national affiliate – which is subject to national policy guidance – may just as properly be regarded as national assets as if they were directly employed in the SPE.

Differences in the role played by organisations of petroleum workers are as great as those among petroleum enterprises. While the right to organise is recognised in virtually all the countries considered here, the degree of concentration in a small number of unions, and of union independence from government or management, varies considerably from one country to another. So does the extent to which unions are concerned with policy issues going beyond wages and working conditions, such as defending the role of the SPE. In some countries workers' organisations play a significant role in identifying and helping meet vocational training needs. Some of the larger and more powerful organisations are found in countries where a large portion of the petroleum sector workforce is concentrated in an SPE.

Table 10. Principles of company and government interaction in petroleum exploration, development and production

Executing entity	Phases					
	Opening	Exploration	Possible commercial discovery	Confirmed commercial discovery	Development	Production
Petroleum company	Preliminary investigations – geological and geophysical Seismic surveys Sea floor and depth surveys, etc. *Collection of environmental technical, social and economic data*	Contracting drilling rig(s) *Determination of mode of exploration/safety* *Establishment of supply bases and administrative centre* Contingency planning Exploratory drilling	Continued exploration and delineation drilling Performance of feasibility studies – Technology – Economy – Production – Prices/markets – Transport *Perform preliminary EIA of development and production phase*	Prepare and submit declaration of commerciality Application for development licence and landing permit Upgrade supply/technology/ *administration* Initiate development planning and design *Initiate contractor, personnel, procurement planning* *Prepare and submit EIS*	Design production facilities (platform/topside/transport/process) *Fabrication and installation* Development drilling Petroleum sales negotiations *Design and fabrication of downstream facilities* *Training for operational phase* *Supplies and services*	*Production, transport, processing of oil and gas* Repairs and maintenance *Retraining* *Monitoring of environmental and socio-economic effects* *Expansion/contraction*

Government (Central/ County/Municipal)	Border delineation *Preparation of legislation and regulations* Determine state participation *Prepare concession/contract system* *Preliminary survey of socio-economic baseline data*	*Concession/contract negotiations* Negotiate participation *Pollution control and contingency regulation* *Commence planning educational/training facilities* *Planning upgrading local industry* *Parliament, cross-sectoral and cross-ministerial consultations* *Perform EIA related to exploration area*	Intensified planning to prepare government sector for rapid decision-making *Prepare development phase* Location of supply bases/ adm. centre/ pipeline landfall/transport/ industry/ infrastructure Instigate education and training Instigate regional planning Instigate local supplies, goods and services Approval of preliminary EIA	*Implementation based on rapid decision-making* *Process parliament application* *Intensify emphasis on training* *Upgrade local industry* *Prepare local/ regional administration* *Safety studies of facilities and equipment* *Approval of EIS* *Parliamentary approval*	*Follow-up on methods, progress, policies* *Procurement monitoring* *Environmental and socio-economic investigation, mitigation, compensation*	*Supervision of contract obligations* *Collection of royalties and taxes*
Implementing level of government agency	Central	Central	Central/County	Central/County/ Municipal	Central/County/ Municipal	Central
Ranges of time requirements	1 to 3 years	1 to 3 years	3 to 6 years	3 to 6 years	3 to 7 years	15 to 25 years

EIA: Environmental and Socio-Economic Impact Assessment
EIS: Environmental and Socio-Economic Impact Statement

Italics identify activities of socio-economic implication.

Source: Norconsult, 1984.

Two further points about institutions are worth making. One is that the countries which have been blessed with stability and professionalism in their petroleum leadership – whether it be in a ministry or an SPE – have fared better, other things being equal, than others where political considerations have dictated frequent personnel changes and/or professionally unqualified appointments. The second, more specific, one is that many countries find it worth considering whether a separate authority for natural gas is needed to promote the efficient development and utilisation of that resource. India has given a positive answer to this question, which has been asked in numerous places because of the perception that people and organisations experienced in oil development are insufficiently sensitive to the special requirements of natural gas.

Passing from the general to the particular, most of the policy issues that arise can be grouped into four principal categories:

Staffing and manpower development

The logic of putting the human resource issues first is that resolving them effectively is a prerequisite for reaping the other benefits of oil and gas development. All countries require certain minimum skills to manage their petroleum industries. These have been identified as expertise in international products supply, in internal distribution and in evaluating the country's geological potential (ILO, 1986, III, pp. 25-26). Where the upstream is concerned, the third of these – and the associated negotiating skills – is of course essential.

As countries from their varied starting-points consider how best to fill the gaps in human skills that may be limiting their upstream development, there is a spectrum of options available. At one end, the least developed countries and those with the most limited or most doubtful geological prospects will find it necessary to rely more heavily on foreign expertise, as available through commercial petroleum companies, through public bodies such as the World Bank, through consulting agencies or through the hiring of individuals on a contract basis. The last of these is perhaps insufficiently used. Where what is required is a knowledge of the operators and procedures common in the industry, and on-the-job training for a small number of counterpart nationals for policy and contract management positions, it can be a cost-effective approach to contract for the services of an experienced person from the industry for a limited period of time. There are available many retirees or other free agents who need not give rise to questions of loyalty, and their compensation even though seemingly high may be a bargain compared with results and to alternative approaches.

The first step is the development of a suitable legal and institutional framework: determination of the policies and terms that will govern petroleum exploration and exploitation and of the government bodies that will be responsible for negotiating with the petroleum companies and/or executing the work themselves. Recruiting, training or contracting for the expertise needed to handle these tasks of policy formulation, legislation and negotiation may be the first human resource issue to be posed.

If exploration is successful, then at an early stage it is advisable to elaborate a flexible and incremental programme for training. Its components may include, in proportions that vary according to the case, sending a small number of qualified nationals abroad for professional training in petroleum-related disciplines; on-the-job training of mid-level local staff by foreign companies operating in the country; and the establishment of new programmes for training additional numbers of skilled workers in crafts needed in, but not restricted to, petroleum applications (see Kohler, 1981; Richter and Duiker, 1984).

For the countries with a good educational infrastructure and the geological basis for expecting substantial production over a long period, more ambitious options may arise. These include the development of specialised petroleum-related programmes in the country's tertiary educational institutions, the establishment of new training institutions for the further development of existing petroleum sector employees, and the formation of research and development institutes to keep abreast of changing technological developments in the industry. Some caution is in order, as it is possible to have too much of a good thing. These are relatively expensive commitments, and if the petroleum activities do not reach or remain at the scale needed to amortise them, they can become an economic and (because of the resulting underemployment of highly skilled people and an associated "brain drain" towards industrial countries) social burden.

Countries benefiting from a relatively strong effort in general education will be best placed to adapt their systems to the petroleum sector's manpower needs. Many of the crafts and the professional skills that the sector requires are not specific to it, and the prior existence of good educational programmes in these basic areas can often be turned rapidly to the advantage of the more specialised needs of the oil and gas sector.

Compatible petroleum production and economic policies

For most NODCs most of the time, the self-sufficiency ratio in hydrocarbons is low, and the main objective is to raise it, towards unity (no net imports or exports) or, if possible, beyond. In these cases, little is required to reconcile petroleum and general economic policies; the macro-economic impact of petroleum production is limited, and generally positive. However, as levels of production approach or surpass those required for self-sufficiency, pressures are generated which may need counteracting if they are not to erode the gains from added petroleum output.

One of these is the tendency for internal demand for petroleum products – stimulated not only by economic and population growth but also by an intangible feeling that less restraint is required in using home-produced than imported oil or gas – to grow so rapidly as to reduce or eliminate the improved trade position that increased petroleum production made possible. In Ecuador, Egypt, Mexico, Pakistan, Tunisia and elsewhere, this process could be seen. The main preventive measures needed are sound pricing policies, which avoid subsidies – except for specified and limited social purposes – in favour of full economic prices.

Second, as the petroleum sector and its revenues become sizeable relative to the rest of the economy, there will be a tendency for some of the negative dynamic effects identified in Chapter 4, such as an overvalued exchange rate, to come into play. In principle, these can be limited by either of two means. Production growth can be restricted as a matter of policy, thus preserving the desired relation (in terms of size) between the petroleum sector and the rest of the economy. Such conservation or depletion policies are subject to a number of practical problems, discussed in Chapter 4, which have limited their application in the past more or less to countries with effective reserves-to-production ratios that are quite high. Hence the interest in the alternative means of limiting the petroleum sector's impact on the rest of the economy: isolating the petroleum revenues in foreign currency-denominated investment accounts, until and as they can be absorbed into the local economy without inflationary or other harmful consequences. In addition to oil-rich countries in the Gulf, such as Kuwait and Saudi Arabia, Trinidad and Tobago used this method in the 1970s in its special funds for long-term development. The practical difficulties involved in this approach concern mainly the fiscal and budgetary, and therefore political, discipline required to carry it through.

Capturing industrial benefits

Giving preference to local suppliers of goods and services to the petroleum industry is a common goal. Contracts with foreign companies frequently specify that they should use local suppliers if possible, subject to given conditions regarding price, quality, etc. There is often a trade-off between the degree of detail in which these commitments are specified and the arrangements for monitoring compliance. The more detailed the requirement, the more mechanical the monitoring process can become. The more general the requirement, the more important it is to have a monitoring and consultative process between government agency and contractor that can assure that good faith and reasonable efforts are made to achieve the objective of optimising the level of local content. In either case, there is no substitute for good communications between the parties, whose interests in these respects need not conflict. Both wish to expedite the production process, and both have an interest in low cost and good quality inputs being available to assist it. It is only on the application of these general principles to specific cases that disagreements may arise, and these instances can be minimised and best handled if the two sides share an information base and an appreciation for each other's objectives.

In a country with a significant degree of industrialisation and a petroleum sector that is relatively small, it may be advisable for the government to help create channels between that sector and potential suppliers so that the latter are aware of the opportunities and requirements. It may be no accident that Brazil and India, which are among the countries that have gone furthest towards supplying the industry's needs from internal sources, both have established mechanisms for this purpose. Joint ventures between local and foreign firms have also been used in several countries – including Brazil, China and India – as an effective vehicle for transferring the technology needed to establish a local production capability (see Mikesell, 1978). Attention needs to be paid also to the need for widely recognised technical standards to be used if locally produced materials are to find ready acceptance with the potential users.

In the case of the very few countries which may aspire to meet the more sophisticated requirements of the industry from internal sources, more active and financially more onerous assistance may be needed, in the form of guarantees, subventions, assured orders, etc., if the industrial capability is to be acquired successfully. There are significant risks in pursuing this course, which should be undertaken only with considerable caution.

The same considerations tend to apply to downstream industrial investments. In contrast to the backward-linked industrial supplies, most of which will be supplied by the private sector if they are locally available and which require only modest government facilitation, the forward-linked projects tend to be with few exceptions large, capital-intensive and associated with sizeable economies of scale. Hence their realisation, especially where they depend on export markets rather than import substitution, tends to require more government underwriting, and involves higher risk and bigger stakes. The rationale for doing such projects can be strong, in countries wishing to diversify their economies and capture more of the value associated with an important natural resource; but rigorous tests should be applied to enhance the likelihood that they will be economically viable over a range of market conditions. The price of wishful thinking, in terms of misallocated resources, can be very high.

Local and regional impacts

One of the more common policy issues associated with petroleum development is how producing areas and their inhabitants can get more benefits than problems from the exploitation of the resource. In almost all instances, this has become an issue; failure to recognise it as such at an early stage almost guarantees that it will become not only an issue but a problem. The tendency of deposits to be found in remote areas, and the high skills requirements of the industry, combine to render difficult the capturing of benefits by the producing area through the means of *direct* employment of the local workforce in the petroleum sector. The employment benefits, especially indirect, are not negligible, and they can be promoted by requiring that preference be given to local candidates, subject to reasonable conditions regarding qualifications.

But the more important policy area concerns the requirements for infrastructure in the area, the provision of resources to meet those requirements, and the relation between the two. Where petroleum exploitation creates new requirements (almost everywhere the case) without additional new resources, local inhabitants will be worse off than before. Where adequate resources and planning are applied to the situation, they may be, and feel, as well or better off than previously.

Experience in several countries shows that a special financial mechanism to allocate an appropriate part of the petroleum revenues (perhaps in the order of 5 per cent) to the areas most affected can be a constructive instrument for dealing with this situation. Other ingredients are also necessary for success. These include effective local and/or regional authorities to administer the funds, planning mechanisms to determine what the needs are and how they can best be met, and sensitive treatment for indigenous populations which may be ill-prepared for the things that a petroleum boom brings in its wake.

* * *

In sum, countries embarking on petroleum production, in expanding their effort beyond historic levels, need to consider a few basic questions at an early point in this process. These include:

- What is the desired size of the petroleum sector relative to the national economy?
- What is the optimum rate of development of upstream activities?
- What should be the relative degrees of reliance on local and foreign firms, persons and equipment at different stages of the process?
- What are the desired roles of the public and private sectors and of market forces versus government-directed activity?
- What government instruments – including means of relating to the private sector – will be most effective in carrying out these choices?
- What consultative channels – with government agencies, employers, workers, representatives of local communities that are affected – will be most effective in forming and maintaining a consensus on these questions?

Events will in any case make it difficult to follow coherent policies; the unforeseen has a way of always happening. But periodic reflection on these fundamental questions will improve the likelihood that desired results will actually be achieved.

Bibliography

Ager-Hanssen, Henrik. 1980. "The exploitation of Norwegian oil and gas", in *Energy Policy*, June 1980.

Allub, Leopoldo; Michel, Marco A. (eds.). 1982. *Impactos regionales de la política petrolera en México*. Mexico D.F., Centro de Investigación para la Integración Social.

American Petroleum Institute. 1983. *Oil exploration in less developed countries: The activities of private oil companies*. Washington, DC, Aug. 1983.

Atta, Jacob K. 1980. "The oil policies of Kuwait: An outsider's interpretation", in *Journal of Energy and Development*, Autumn, 1980.

Baum, Vladimir. 1985. "Thailand: Diversifying energy supplies", in *Petroleum Economist*, Feb. 1985.

Bennett, Amanda. 1984. "Xenophobia still tints China's outlook", in *Wall Street Journal*, 23 Aug. 1984.

Bienen, Henry. 1983. *Oil revenues and policy choice in Nigeria*. World Bank Staff Working Papers No. 592, Washington, DC.

Blitzer, Charles R. et al. 1983. *Overcoming the oil exploration gap in developing countries*. Massachusetts Institute of Technology Center for Energy Policy Research, for the United States Agency for International Development, Oct. 1983.

Bourcier, Philippe. 1984. "Politique de mise en valeur du gaz naturel dans les pays en voie de développement", in *Revue de l'énergie*, Aug.-Sep. 1984.

Bowie, Paddy. 1985. *Petroleum and development in Malaysia*. Unpublished paper, International Labour Office.

BP Statistical Review of World Energy. 1984. British Petroleum Company plc, London.

Brennand, T. P. 1985. *The role of the private sector in petroleum exploration*. United Nations Symposium on Financing of Petroleum Exploration and Development in Developing Countries, Athens, Greece, April 1985.

Broadman, Harry G. 1985. *An econometric analysis of the determinants of exploration for petroleum outside North America*. Washington, DC, Resources for the Future.

Cairns, William J.; Rogers, Patrick M. (eds.). 1981. *Onshore impacts of offshore oil*. London, Applied Science Publishers.

Carley, Michael J.; Edward S. Bustelow. 1984. *Social impact assessment monitoring: A guide to the literature*. Boulder, Colorado, Westview.

Chulalongkorn University, Faculty of Economics. 1985. *Economic impacts of Sirikit oil field*. Bangkok.

Collier, Paul. 1983. "Oil and inequality in rural Nigeria", in Dharam Ghai and Samir Radwan (eds.). *Agrarian policies and rural poverty in Africa*. Geneva, ILO.

Commander, Simon; Peek, Peter. 1983. *Oil exports, agrarian change and the rural labour process: The Ecuadorian Sierra in the 1970s* (Geneva, ILO; mimeographed World Employment Programme research working paper; restricted).

Coopers & Lybrand Associates. 1982. *Eastern Seaboard Study, 1982.*

Dasgupta, P.; Heal, G.M. 1974. "The optimal depletion of exhaustible resources", in *Review of Economic Studies*. Symposium on the Economics of Exhaustible Resources.

"Ecuador: CEPE", in *OPEC Bulletin*, Feb. 1985.

Economist Intelligence Unit. 1984 and 1985. *Quarterly Energy Review*, Latin America and the Caribbean. No. 2, 1984; and No. 2, 1985.

El Mallakh, Ragaei; Oystein, Noreng; Poulson, Barry W. 1984. *Petroleum and economic development: The cases of Mexico and Norway*. Lexington and Toronto, Lexington.

Elf Gabon 1949-1979. 1979. Paris, Elf Aquitaine.

Energy Information Administration. 1983. *The petroleum resources of Venezuela and Trinidad and Tobago*. Washington, DC.

Engelshoven, J.M.H. van. 1985. "Responses to change in the energy scene". Speech delivered to the Conference on Global Oil Policy Responses, Lima, Peru, 27 February 1985. London, Shell International Petroleum Company.

Environmental planning for large-scale development projects. 1983. Final Report of the International Workshop on Environmental Planning for Large-Scale Development Projects, Whistler, British Columbia.

European Chemical News, 25 Mar. 1985.

Evans, Margaret I. 1984. *Firewood versus alternatives: Domestic fuel in Mexico*. Oxford, Commonwealth Forestry Institute.

Exxon Corporation. 1985. *Development of oil resources in oil importing developing countries*. Unpublished paper, Feb. 1985.

—. *This is Exxon*. New York, 1985.

Favre, Jean; Le Leuch, F.M. 1982. "Petroleum exploration trends in the developing countries", in United Nations: *Petroleum exploration strategies in developing countries*. London, Graham and Trotman.

Financial Times (London), 8 July 1985.

Foster, John. 1985. "Petroleum exploration and development trends in the developing countries". Paper presented at the United Nations Symposium on Financing of Petroleum Exploration and Development in Developing Countries, Athens, Greece, April 1985.

Gall, Norman. 1985. "Petroleum and economic development in Brazil". Unpublished paper.

Gallagher, Charles F. 1980. "Population, petroleum, and politics: Mexico at the crossroads". American Universities Field Staff Reports, No. 42.

Gault, John. 1983. "Decision making for natural gas development", in *Journal of Energy Development*, spring, 1983.

— (ed.). 1985. *The economics of natural gas in developing countries*. Oxford, Pergamon.

Gelb, Alan H. 1981. "Capital-importing oil exporters: Adjustment issues and policy choices". World Bank Staff Working Paper No. 475. Washington, DC, Aug. 1981.

Gilbert, John T.E. (ed.). 1983. *Technical guidelines for offshore oil and gas development.* Tulsa, PennWell.

Grayson, George. 1981. *The politics of Mexican oil.* Pittsburgh, University of Pittsburgh Press.

Grossling, Bernardo F.; Neilsen, Diane T. 1985. *In search of oil.* London, Financial Times Business Information.

Haan, T.J. 1985. *De oliesector en de overheidsfinancien van de Nederlandse Antillen (The oil sector and public finance of the Dutch Antilles).* Rotterdam, ESB.

Hartshorn, J.E. 1985. "Natural gas development begins at home", in *Energy.* Feb. 1985.

Hills, Peter. 1985. "The development of the petroleum industry in the People's Republic of China". Unpublished paper. ILO.

IEDC/IMI Symposium on Petroleum Exploration and Production in Developing Countries. 1983. Geneva, International Energy Development Corporation.

International Energy Agency. 1984. *Energy balances of developing countries, 1971/1982.* Paris, OECD.

International Labour Office. 1981. *Multinationals' training practices and development.* Geneva.

—. 1978. *Safety problems in the offshore petroleum industry.* Geneva.

—. 1977. *Social and labour practices of multinational enterprises in the petroleum industry.* Geneva.

International Labour Organisation, Chemical Industries Committee, Ninth Session. 1982. *General report.* Geneva.

—. 1986. Petroleum Committee, Tenth Session. I. *General report.* II. *Occupational safety and health and the working environment in the petroleum industry.* III. *Manpower planning and development in the petroleum industry.* Geneva.

Jones, Aubrey. 1981. *Oil: The missed opportunity, or Naft and Shaft.* London, André Deutsch.

Joseph, Richard A. 1978. "Affluence and underdevelopment: The Nigerian experience", in *The Journal of Modern African Studies.* 1978. No. 2.

Kambara, T. 1984. "China's energy development during the readjustment and prospects for the future", in *China Quarterly,* Dec. 1984.

Khan, Kameel I.F. "The legal and administrative regulation of the petroleum industry in Trinidad and Tobago: A study of law and policy in petroleum development". Thesis, Institute of Comparative Law, McGill University, Montreal.

Khitoun, A.D. 1985. *Social and economic effects of petroleum and gas development in Western Siberia.* Unpublished paper. ILO.

Kohler, L.R. 1981. *Meeting the Arab petroleum industry's manpower needs: Selected issues.* Paper presented to the APTI Seminar on Problems and Sufficiency of Manpower in Arab Petroleum Industry, Baghdad.

Lanfranco, Sam. 1981. "Mexican oil, export-led development and agricultural neglect". Working Paper No. 1, Centre for Research on Latin America and the Caribbean, York University, Feb. 1981.

Lerat, Serge. 1984. "Le pétrole en Chine", in *Cahiers d'Outre-Mer,* Oct.-Dec. 1984.

Lesch, Ann Mosely. 1985. "Rebellion in the Southern Sudan", in *UFSI Reports,* No. 8.

Mackay, K.H. 1983. "Estimates of manpower requirements and manpower availability in the world petroleum industry, 1980-2000, and their implications". *Proceedings of the Eleventh World Petroleum Congress, Vol. V.* Chichester, Wiley.

Madujibeya, S.A. 1976. "Oil and Nigeria's economic development", in *African Affairs*, July 1976.

Marchán C., Cornelio. 1985. *Algunas apreciaciones sobre los efectos económicos y sociales del desarrollo petrolero en el Ecuador*. Unpublished paper. ILO.

Mayorga A., Eleodor. 1985. *The social and economic effects of petroleum development in Peru.* Unpublished paper. ILO.

Mendoza Berrueto, Eliseo. 1984. "Mexico's strategy meets the changing world oil market", in *OPEC Bulletin*, Apr. 1984.

Mikesell Raymond. 1978. "International transfer of technology in the extractive industries", in *Public policy and technology transfer: Viewpoints of US business*, Vol. II, *Industry characteristics*. New York.

—. 1984. *Petroleum company operations and agreements in the developing countries.* Washington, DC, Resources for the Future.

National Petroleum Council. 1982. *Third World petroleum development: A statement of principles*. Washington, DC.

Netherlands Economic Institute. 1984. *The significance of Shell investments for the Dutch economy.* Rotterdam.

New Straits Times. Kuala Lumpur, 14 June 1985.

Norconsult, 1983a. Desk Study. "An oil economy striving for balanced development – The case of Nigeria", Dec. 1983.

—. 1983b. Desk Study. "Petroleum and socio-economic development in Mexico", Oct. 1983.

—. 1983c. *Mission Report No. 2: South America*. Oslo, May 1983.

—. 1983d. *Mission Report No. 3: Africa*. Oslo, Apr. 1983.

—. 1984. *Social and economic effects of petroleum development programmes*. Geneva.

Oil and Gas Journal, 31 Dec. 1984 and 6 May 1985.

Onoh, J.K. 1983. *The Nigerian oil economy: From prosperity to glut*. London, Croom Helm.

OPEC Bulletin. 1985. "Country profile: Pakistan", Feb. 1985.

PA Management Consultants. 1985. "Manpower planning in the petroleum industry". Amsterdam.

Pachauri, R.K. 1977. *Energy and development in India*. New York, Praeger.

Parra, A.A. 1977. "Manpower planning and development in Petroleos de Venezuela", in *Proceedings of the OPEC Seminar on the Present and Future Role of the National Oil Companies*, Vienna.

Parra, Francisco. 1985. *Problems and prospects for accelerated petroleum exploration in the developing world.* United Nations Symposium.

Petroleum Authority of Thailand. *Gas plant project.*

Petroleum Economist. Various issues.

Planning Commission, Government of Pakistan. 1983. *The Sixth Five Year Plan.* Islamabad, Oct. 1983.

Prevot-Schapira, Marie-France. 1982. "Trabajadores del petróleo y poder sindical en México", in M.S. Wionczek (ed.). *Energía en México: ensayos sobre el pasado y el presente.* Mexico D.F., El Colegio de México.

Randall, Laura. Forthcoming study on economic effects of petroleum development in Brazil, Mexico and Venezuela.

Richter, L.; Duiker, P. 1984. *Manpower and training assessments for petroleum development in the Arab region.* Paper presented to the APTI meeting for Training Directors and Training Senior Staff in the Oil Sector in OAPEC Member States, Tunis.

Sabra, Yasser. 1985. "Les effets sociaux et économiques de la production du pétrole en Egypte". Unpublished paper. ILO.

Sanders, Thomas G. 1982. "Oil and the Mexican economy", in *UFSI Reports*, No. 12.

Saramma, A.D. 1985. "Socio-economic effects of oil and gas from Bombay High". Unpublished paper. ILO.

Schliephake, Konrad. 1977. *Oil and regional development: Examples from Algeria and Tunisia.* New York, Praeger.

Seers, Dudley. 1978. *The life cycle of a petroleum economy.* Institute of Development Studies (University of Sussex) Discussion Paper. Brighton, Sep. 1978.

—. 1969. "A step towards a political economy of development: Trinidad/Tobago", in *Social and Economic Studies*, Sep. 1969.

Shell Internationale Petroleum Maatschappij, Group Personnel. 1980. "Training in the Royal Dutch/Shell Group of Companies". The Hague.

Société nationale Elf Aquitaine. 1983. *Rapport social.* Paris.

Spurrier, Walter R. (ed.). 1984. *Weekly Analysis of Ecuadorean Issues.* Quito, 24 Sep. 1984.

Stauffer, Thomas R. (with Frank H. Lennox). 1984. *Accounting for "wasting assets": Income measurement for oil and mineral-exporting rentier States.* Vienna, OPEC Fund for International Development.

Sterner, Thomas. 1982. "Economic effects of the oil expansion in Mexico". Memorandum No. 83. University of Gothenburg, Department of Economics.

Survival International Bulletin and *News.* Various issues.

Swepston, Lee; Plant, Roger. 1985. "International standards and the protection of the land rights of indigenous and tribal populations", in *International Labour Review*, Jan.-Feb. 1985.

Tata Energy Research Institute with the assistance of C.R. Jagannathan. 1985. *Social and economic aspects of the oil and gas industry in India.* Unpublished paper. ILO.

Tingsabadh, Charit. 1985. *Social and economic effects of petroleum development programme in Thailand.* Unpublished paper. ILO.

Turner, Terisa. 1982. *Multinational enterprises and employment in the Caribbean with special reference to Trinidad and Tobago.* Multinational Enterprises Programme Working Paper No. 20. ILO.

United Nations, Centre for Natural Resources, Energy, and Transport. 1980. *State petroleum enterprises in developing countries.* New York, Pergamon.

—. 1983. Centre on Transnational Corporations. *Main features and trends in petroleum and mining agreements*. New York.

—. 1985. Committee on Natural Resources. *Energy resources: Trends and salient issues*; E/C.7/1985/4, 7 Mar. 1985.

United Nations Conference on Trade and Development. 1984. *Technological problems in the energy field in the developing countries: Development of the technological capability of the petroleum industry in Peru;* UNCTAD/TT/59, Dec. 1984.

United Nations. 1984. *Development and international economic co-operation: Development of the energy resources of developing countries;* A/39/420, 24 Aug. 1984.

United Nations Environment Programme (UNEP). 1980. *Guidelines for assessing industrial environmental impact and environmental criteria for the siting of industry*. Paris.

United Nations Industrial Development Organisation (UNIDO). 1978. "First world-wide study on the petrochemical industry: 1975-2000". Vienna.

United Nations Symposium on Financing of Petroleum Exploration and Development in Developing Countries. 1985. Athens, Greece, April.

Velasco-S.; Jesus-Agustin. 1983. *Impacts of Mexican oil policy on economic and political development*. Lexington, Mass., Lexington.

Von Lazar, Arpad; Magid, Bruce. 1975. "Energy policy and social development: Notes on Trinidad-Tobago and Venezuela", in *Energy Policy*, Sep. 1975.

Wionczek, Miguel S. 1982. "Mexico's energy policy – The past and the future", in *Journal of Energy and Development*, Spring, 1982.

—. *Ivory Coast: Issues and options in the energy sector*. Report No. 5250-IVC (restricted). Washington, DC, 1985.

World Bank. 1984a. *Ecuador: An agenda for recovery and sustained growth*. Report No. 5094-EC (restricted). Washington, DC, Oct. 1984.

—. 1984b. *Energy issues and options in thirty developing countries*. Report No. 5230 (restricted). Washington, DC.

—. 1984c. *Tanzania: Issues and options in the energy sector*. Report No. 4969-TA (restricted). Washington, DC.

—. 1983a. *The energy transition in developing countries*. Washington, DC.

—. 1983b. *Trinidad and Tobago: Development issues for the 1980s*. Report No. 4202-TR (restricted). June 1983.

—. *Mexico oil impact on regional development: A case study of the state of Tabasco*. Report No. 3087-ME (restricted). Washington, DC, 1980.

—. Energy Department. 1983. *Marginal cost of natural gas in developing countries: Concepts and applications*. Washington, DC, 1983.

Yamaguchi, S.; Kuczek, E. 1984. "The social and economic impact of large-scale energy projects on the local community", in *International Labour Review*, Mar.-Apr. 1984.

Zakariya, Hasan S. 1982. "Transfer of technology under petroleum development contracts", in *Journal of World Trade Law*, May-June 1982.

Zymelman, M. 1980. *Forecasting manpower demand and occupational structures of industries*. Washington, DC, World Bank.